EAI/Springer Innovations in Communication and Computing

Series Editor

Imrich Chlamtac, European Alliance for Innovation, Ghent, Belgium

The impact of information technologies is creating a new world yet not fully understood. The extent and speed of economic, life style and social changes already perceived in everyday life is hard to estimate without understanding the technological driving forces behind it. This series presents contributed volumes featuring the latest research and development in the various information engineering technologies that play a key role in this process. The range of topics, focusing primarily on communications and computing engineering include, but are not limited to, wireless networks; mobile communication; design and learning; gaming; interaction; e-health and pervasive healthcare; energy management; smart grids; internet of things; cognitive radio networks; computation; cloud computing; ubiquitous connectivity, and in mode general smart living, smart cities, Internet of Things and more. The series publishes a combination of expanded papers selected from hosted and sponsored European Alliance for Innovation (EAI) conferences that present cutting edge, global research as well as provide new perspectives on traditional related engineering fields. This content, complemented with open calls for contribution of book titles and individual chapters, together maintain Springer's and EAI's high standards of academic excellence. The audience for the books consists of researchers, industry professionals, advanced level students as well as practitioners in related fields of activity include information and communication specialists, security experts, economists, urban planners, doctors, and in general representatives in all those walks of life affected ad contributing to the information revolution.

Indexing: This series is indexed in Scopus, Ei Compendex, and zbMATH.

About EAI - EAI is a grassroots member organization initiated through cooperation between businesses, public, private and government organizations to address the global challenges of Europe's future competitiveness and link the European Research community with its counterparts around the globe. EAI reaches out to hundreds of thousands of individual subscribers on all continents and collaborates with an institutional member base including Fortune 500 companies, government organizations, and educational institutions, provide a free research and innovation platform. Through its open free membership model EAI promotes a new research and innovation culture based on collaboration, connectivity and recognition of excellence by community.

Ahmed A. Abd El-Latif • Lo'ai Tawalbeh •
Yassine Maleh • Brij B. Gupta
Editors

Secure Edge and Fog Computing Enabled AI for IoT and Smart Cities

Includes selected Papers from International Conference on Advanced Computing & Next-Generation Communication (ICACNGC 2022)

Springer

EAI
RESEARCH MEETS INNOVATION

Editors
Ahmed A. Abd El-Latif
Center of Excellence in Quantum and Intelligent Computing
Prince Sultan University
Riyadh, Saudi Arabia

Department of Mathematics and Computer Science, Faculty of Science
Menoufia University
Shebin El-Koom, Egypt

Yassine Maleh
Sultan Moulay Slimane University
Settat, Morocco

Lo'ai Tawalbeh
Texas A&M University
San Antonio, TX, USA

Brij B. Gupta
Department of Electrical and Computer Engineering
Lebanese American University
Beirut, Lebanon

Department of Computer Science and Information Engineering
Asia University
Taichung, Taiwan

ISSN 2522-8595 ISSN 2522-8609 (electronic)
EAI/Springer Innovations in Communication and Computing
ISBN 978-3-031-51096-0 ISBN 978-3-031-51097-7 (eBook)
https://doi.org/10.1007/978-3-031-51097-7

This Springer imprint is published by the registered company Springer Nature Switzerland AG
The registered company address is: Gewerbestrasse 11, 6330 Cham, Switzerland

Preface

Edge/fog computing is considered as a new approach with the potential to impact our daily life and the world economy profoundly. The edge server layer is expected to be a medium that intelligently optimizes the use of computing resources. This technology innovation has begun to drive new levels of performance and productivity in multiple domains. Meanwhile, cloud computing is becoming a major enabler of various industrial innovations. Within the transformations, edge/fog computing is a quick-powered component that combines cloud computing with other network-based techniques, such as Internet of Things (IoT) and web services. Even though edge/fog computing has great potential to speed up the next generation of networking implementations, security and privacy concerns still restrict its applications. The threats exist at different layers because more parties are involved in the service processes. Both detecting threats and finding out solutions are significant. Building up an intelligent security mechanism in edge/fog computing is an urgent mission in an advanced network setting.

This book brings recent research in security and privacy topics to discuss, evaluate, and improve the novel approaches to data protection in IoT and edge/fog computing. The primary focus of this book will address novel security mechanisms in IoT and edge/fog computing, advanced secure deployments for large scaled edge/fog computing, intelligent secure solutions, and new efficient data security strategy for IoT and edge/fog computing. Note: Papers, after rigorous revisions, are selected from the International Conference on Advanced Computing and Next-Generation Communication (https://icacnc.com/).

This book provides a relevant reference for students, researchers, engineers, and professionals working in this particular area or those interested in grasping its diverse facets and exploring the latest advances in secure edge and fog computing enabled AI for IoT and smart cities.

We would like to sincerely thank the authors of the contributing papers as well as reviewers for their valuable suggestions and feedback. The editors would like to thank Dr. Eliska Vlckova (Springer, Managing Editor at EAI – European Alliance for Innovation) for the editorial assistance and support to produce this important scientific work. Without this collective effort, this book would not have been possible to be completed.

We hope you will enjoy this book and this amazing research field of cybersecurity and next-gen-enabled AI for IoT and smart cities.

Riyadh, Saudi Arabia — Ahmed A. Abd El-Latif
San Antonio, TX, USA — Lo'ai Tawalbeh
Settat, Morocco — Yassine Maleh
Taichung, Taiwan — Brij B. Gupta

Introduction

With the advancement and deployment of Internet of Things (IoTs) in smart cities, a growing number of physical objects are being connected by IoTs to provide services to users [1, 2]. It is the Internet working of various objects and network connectivity that allows these objects to communicate and exchange data, including sensors, smart meters, smart phones, smart vehicles, radio-frequency identification tags, personal digital assistants, and other items (embedded with electronics, software and actuators). The interconnection of these devices enables advanced IoT applications, e.g., product tracking, environment monitoring, patients' surveillance, and energy management, and expands the automation to our daily life [3]. Some IoT applications might be latency-sensitive; some might involve private data, which should be stored and processed locally; and some might produce large volumes of data, which could be a heavy burden for networks [2]. Moreover, an increasing number of devices (e.g., smart glasses, smart phones, and vehicles) are involved in IoT for collecting and delivering fine-grained data, which may contain multimedia information (e.g., photos, videos, and voices). The large amounts of data result in heavy network congestion and complicated processing load on devices and control systems.

Edge computing (EC) and fog computing have been integrated to address the above issues in IoT, which provide resources at the user end in a decentralized manner and respond to end users faster than the cloud computing [1]. It can enable analytic and knowledge generation to occur at the source of the data close to user end. It leverages resources that may not be continuously connected to the network such as laptops, smart phones, tablets and sensors. As a nontrivial extension of cloud computing, it is inevitable that some issues will continue to persist, especially security and privacy issues [4]. Fog computing is deployed by different fog service providers that may not be fully trusted and devices are vulnerable to be compromised. Thus, EC servers are confronted with various security and privacy threats [5].

The IoT have constrained computing, storage, and battery resources and are easy to be hacked, broken, or stolen. Although the existing solutions in cloud computing could be migrated to address some security and privacy issues in EC, it still has its specific security and privacy challenges due to its distinctive features, such as decentralized infrastructure, mobility support, location awareness, and low latency. In addition, EC also offers a more secure infrastructure than cloud computing because of the local data storage and the non-real-time data exchange with cloud centers. EC servers could be represented as proxies for end-devices to perform secure operations, if the devices lack sufficient resources to do so [6, 7]. Unfortunately, security and privacy issues and security resources in EC have not been systematically identified.

Artificial intelligence (AI) exhibits great potential in addressing security issues in EC/FC, which attracts the attention of researchers from both academia and industry. AI plays a pivotal role in identifying and mitigating security threats. AI (machine learning, deep learning, etc.) algorithms can be deployed on edge devices to continuously monitor and assess the integrity of the devices and their data. This enables proactive threat detection and rapid response to security incidents, thus enhancing the overall security posture of IoT in smart cities [8].

Structure of the Book

In this book, we have organized the content into *two parts as* shown in Table 1: Part I is AI-Enabled Smart City IoT System using Edge /Fog Computing, and Part II is about AI-Enabled Smart City IoT System Using Edge /Fog Computing. This book offers a comprehensive perspective on leveraging cutting-edge technologies to create smarter and more secure urban environments. It spans from innovative applications in Part I to the essential security considerations in Part II, providing valuable insights for researchers, practitioners, and policymakers in the field of smart city development.

About This Book

The 19 full papers presented in this book were carefully reviewed and selected from 112 submissions. The book covers a wide range of topics in the areas of security and AI for IoT and smart cities: artificial intelligence, federated learning, deep learning, edge computing, fog computing, UAVs, blockchain, cybersecurity, etc.

About Part I: Papers from 1 to 11 in Part I covers a variety of research topics and technologies in AI-enabled smart city IoT system using edge/fog computing. The list of topics in this part is a collection of specific applications, technologies, and research areas that can benefit from the broader field of AI-enabled smart city IoT systems and edge/fog computing, and each contributes to the overall exploration of the field within the book.

Table 1 Taxonomy of the topics in the book

Parts	Part I: AI-Enabled Smart City IoT System Using Edge /Fog Computing	Part II: Fog/Edge Computing Security Issues
Theme	This part focuses on AI-enabled smart city IoT systems, with a special emphasis on the utilization of edge and fog computing technologies. This section brings together a collection of cutting-edge research and applications in the domain of smart cities.	This part focuses on the critical theme of fog/edge computing security issues within the context of IoT, smart cities, and cloud computing. It delves into various aspects of security and data privacy, shedding light on cutting-edge techniques and solutions. The chapters within this section explore the following key topics: secure edge computing, big data analytics, access and trust management, blockchain, authentication, etc.
Papers	1. Multi-level edge computing system for autonomous vehicles 2. UAVs-based edge computing system for smart city applications 3. Organization of smart city services based on microservice architecture 4. Pseudo-random error-correcting codes in network coding 5. Proactive management in smart city: transport convoys 6. Federated learning for linux malware detection: an Experimental Study 7. Delay prediction in M2M networks using deep learning approach 8. Energy-efficient beam shaping in MIMO system using machine learning 9. Channel cluster configuration selection method for IEEE 802.11 networks planning 10. Service migration algorithm for UAVs recharge zones in future 6G network 11. FedBA: Non-IID federated learning framework in UAV networks	12. Big data analytics for secure edge-based manufacturing Internet of Things (MIoT) 13. Artificial intelligence-based secure edge computing systems for IoTDs and smart cities: A survey 14. Machine learning techniques for secure edge SDN 15. Machine learning-based identity and access management for cloud security 16. Spatial data of smart cities: Trust 17. Smart city infrastructure projects: Spatial data of risks 18. A comparative analysis of blockchain-based authentication models for IoT networks 19. Development of determining a wireless client location method in the IEEE 802.11 network in order to ensure the IT infrastructure security

About Part II: Papers from 12 to 19 in Part II provide a comprehensive exploration of security challenges and solutions in fog and edge computing, making it a valuable resource for researchers, practitioners, and policymakers in the field of IoT, smart cities, and cloud computing.

Key Aspects of the Book

The book is a comprehensive exploration of cutting-edge technologies and solutions aimed at transforming urban environments into smarter and more efficient spaces. It is divided into two distinct parts, each addressing critical aspects of AI-enabled smart city IoT systems using edge/fog computing. The book introduces various techniques, such as secure edge and fog computing–enabled AI for IoT and smart cities. It delves into the following key aspects:

- Discussion, evaluation, and enhancement of approaches for data protection in IoT and edge/fog computing.
- Establishment of a solid foundation in the core concepts and principles of IoT and 5G security in the context of edge/fog computing.
- Inclusion of selected papers from ICACNGC 2022: International Conference on Advanced Computing and Next-Generation Communication, Saint Petersburg, Russia, October 12–13, 2022.

Target Audience

The book is intended for researchers, academics, engineers, and professionals in the field of secure edge and fog computing–enabled AI for IoT and smart cities. The book will also be useful for graduate and postgraduate students who are interested in these areas.

Saint Petersburg, Russia
Riyadh, Saudi Arabia
Menoufia, Egypt

Ahmed A. Abd El-Latif

References

1. Asim, M., Wang, Y., Wang, K., Huang, P.Q.: A review on computational intelligence techniques in cloud and edge computing. IEEE Trans. Emerg. Topics Comput. Intell. **4**(6), 742–763 (2020)
2. Shi, W., Cao, J., Zhang, Q., Li, Y., Xu, L.: Edge computing: vision and challenges. IEEE Internet of Things J. **3**(5), 637–646 (Oct 2016)
3. Al-Fuqaha, A., Guizani, M., Mohammadi, M., Aledhari, M., Ayyash, M.: Internet of things: a survey on enabling technologies, protocols, and applications. IEEE Commun. Surv. Tutor. **17**(4), 2347–2376 (2015)
4. Ali, K.A., Vinod, K., Musheer, A., Gupta, B.B., Musheer, A., Abd El-Latif, A.A.: A secure and efficient key agreement framework for critical energy infrastructure using mobile device. Telecommun. Syst. **7**, e700 (2021)

5. Abd El-Latif, A.A., Abd-El-Atty, B., Venegas-Andraca, S., Mazurczyk, W.: Efficient quantum-based security protocols for information sharing and data protection in 5g networks. Future Gener. Comput. Syst. **100**, 5 (2019)
6. Kaur, M., Singh, D., Kumar, V., Gupta, B.B., Abd El-Latif, A.A.: Secure and energy efficientbased e-health care framework for green internet of things. IEEE Trans. Green Commun. Netw., **5**(3):1223–1231 (2021)
7. Abd El-Latif, A.A., Abd-El-Atty, B., Mehmood, I., Muhammad, K., Venegas Andraca, S.E., Peng, J.: Quantum-inspired blockchain-based cybersecurity: securing smart edge utilities in IOT-based smart cities. Inf. Process. Manag. **58**(4), 102549 (2021)
8. Masud, M., Gaba, G.S., Choudhary, K., Hossain, M.S., Alhamid, M.F., Muhammad, G.: Lightweight and anonymity-preserving user authentication scheme for IoT-based healthcare. IEEE Internet of Things J. **9**(4), 2649–2656 (15 Feb 2022)

Contents

About the Editors

Ahmed A. Abd El-Latif graduated with distinction from the Harbin Institute of Technology, China, in 2013, earning a PhD. Since then, he has led and participated in several successful research projects and secured grants in Egypt, the Russian Federation, Saudi Arabia, China, Malaysia, and Tunisia. Currently, Dr. Abd El-Latif holds staff positions at Menoufia University, Egypt, and Prince Sultan University, Saudi Arabia. With over 18 years of professional experience, he has published approximately 300 papers in journals and conference proceedings, including 12 books, with over 10,000 citations. Since 2022, he has served as head of the MEGANET 6G Lab Research in the Russian Federation. Dr. Abd El-Latif holds several leadership positions, including vice chair of the EIAS Research Lab and founder and deputy director of the Center of Excellence in Quantum and Intelligent Computing (Prince Sultan University, Saudi Arabia). He has received several awards, including the State Encouragement Award in Engineering Sciences from the Arab Republic of Egypt in 2016, the Best PhD Student Award from the Harbin Institute of Technology, China, in 2013, and the Young Scientist Award from Menoufia University, Egypt, in 2014. Dr. Abd El-Latif actively participates in the scientific community, serving as chair/co-chair of several Scopus/EI conferences. He also holds editorial positions, including editor-in-chief of the *International Journal of Information Security and Privacy*, series editor of Quantum Information Processing and Computing, and series editor of Advances in Cybersecurity Management. Additionally, he serves as an academic editor or associate editor for many indexed journals in the Web of Science (WoS) and Scopus, covering various quartiles. Dr. Abd El-Latif's research interests span quantum communications and cryptography, cybersecurity, artificial intelligence of things (AIoT), AI-based image processing, information hiding, and applications of dynamical systems (discrete-time models: chaotic systems and quantum walks) in cybersecurity.

Yassine Maleh is an associate professor of cybersecurity and IT governance at Sultan Moulay Slimane University, Morocco. He is a PhD in computer sciences, since 2017. He is the founding chair of IEEE Consultant Network Morocco and founding president of the African Research Center of Information Technology and Cybersecurity. He is a senior member of IEEE and a member of the International Association of Engineers IAENG and the Machine Intelligence Research Labs. He has made contributions in the fields of information security and privacy, Internet of Things security, and wireless and constrained networks security. His research interests include information security and privacy, Internet of Things, networks security, information system, and IT governance. He has published over than 100 papers (book chapters, international journals, and conferences/workshops), 17 edited books, and 3 authored books. He is the editor-in-chief of the *International Journal of Information Security and Privacy* and the *International Journal of Smart Security Technologies* (IJSST). He serves as an associate editor for *IEEE Access* (2019 Impact Factor 4.098), the *International Journal of Digital Crime and Forensics* (IJDCF), and the *International Journal of Information Security and Privacy* (IJISP). He is a series editor of *Advances in Cybersecurity Management*, by CRC Taylor & Francis?-. He was also a guest editor of a special issue on *Recent Advances on Cyber Security and Privacy for Cloud-of-Things of the International Journal of Digital Crime and Forensics* (IJDCF), Volume 10, Issue 3, July–September 2019. He has served and continues to serve on executive and technical program committees and as a reviewer of numerous international conferences and journals, such as *Elsevier Ad Hoc Networks*, *IEEE Network Magazine*, *IEEE Sensor Journal*, *ICT Express*, and *Springer Cluster Computing*. He was the publicity chair of BCCA 2019 and the general chair of the MLBDACP 19 symposium and ICI2C'21 Conference. He received Publons Top 1% Reviewer Award for the years 2018 and 2019.

Brij B. Gupta completed his PhD in Computer Science and Engineering from the Indian Institute of Technology Roorkee, India, in 2011. With over 18 years of professional experience, he has authored over 500 papers in journals and conferences, including 35 books and 12 patents, with over 27,000 citations. Prof. Gupta has received numerous national and international awards, including the Canadian Commonwealth Scholarship (2009), Faculty Research Fellowship Award (2017) from MeitY, Government of India, IEEE GCCE Outstanding and WIE Paper Awards, and the Best Faculty Award (2018 and 2019) from NIT KKR, respectively. He has been recognized as a Clarivate Web of Science Highly Cited Researcher in Computer Science (top 0.1% researchers globally) consecutively in 2022 and 2023 and has been included in Stanford University's ranking of the world's top 2% scientists consecutively in 2020, 2021, 2022, and 2023. Prof. Gupta is also a visiting/adjunct professor at several universities worldwide. He became an IEEE senior member in 2017 and was selected as the 2021 distinguished lecturer in IEEE CTSoc. Dr. Gupta serves as a member-in-large on the Board of Governors of the IEEE Consumer Technology Society (2022–2024). He currently serves as the editor-in-chief of *IJSWIS*, *IJSSCI*, *STE*, and *IJCAC* and leads a book series with CRC and IET press. He has also served as a TPC member for over 150 international conferences and as an associate/guest editor for various journals and transactions. Prof. Gupta serves as the director of the International Center for AI and Cyber Security Research and Innovations (CCRI) and holds the position of distinguished professor in the Department of Computer Science and Information Engineering (CSIE) at Asia University, Taiwan. His research interests include information security, cyber-physical systems, cloud computing, blockchain technologies, intrusion detection, AI, social media, and networking.

Lo'ai Tawalbeh (IEEE SM) completed his PhD degree in Electrical and Computer Engineering from Oregon State University in 2004 and MSc in 2002 from the same university with GPA 4/4. Dr. Tawalbeh is currently an associate professor at the Department of Computing and Cybersecurity at Texas A&M University-San Antonio. Before that, he was a visiting researcher at the University of California-Santa Barbara. Since 2005, he taught/developed more than 25 courses in different disciplines of computer engineering and science with focus on cybersecurity for the undergraduate/graduate programs at the New York Institute of Technology (NYIT), DePaul's University, and Jordan University of Science and Technology. Dr. Tawalbeh won many research grants and awards with over than 2 million USD. He has over 80 research publications in refereed international journals and conferences.

Part I
AI Enabled Smart City IoT System Using Edge/Fog Computing

Multilevel Edge Computing System for Autonomous Vehicles

Mohammed Saleh Ali Muthanna, Dmitry Elkin, Semyon Likhtin, and A. M. Al-Sveiti Malik

1 Introduction

In recent years, vehicle-to-everything (V2X) communication has attracted significant research interest by both academia and industry. It has a voluminous range of wireless technologies such as vehicle-to-vehicle (V2V) car communication with each other, vehicle-to-infrastructure (V2I) communication of the car, and infrastructure, vehicle-to-pedestrian (V2P) vehicle-to-pedestrian (V2P) vehicle-to-pedestrian or cyclist communication, among others. 5G New Radio (5G NR) V2X technology was developed to solve this problem. Also, in issue 15, 3rd Generation Partnership Project (3GPP) improved the performance of the PC5 interface (LTE-eV2X) [1–4].

However, a rapid increase in the number of cars is expected, and this will lead to a large increase in the number of network devices. Also, these devices will have to support new features, from smoother screen movement to even holographic images. This is the amount of data from a high-density network. Data transmission must be transmitted with ultralow latency of 1 Ms. This level of characteristics cannot be supported based on 5G networks. The solution to this problem will be using 6G technology in such a network.

Intermediate servers for computing between the cloud and the endpoints are also required to ensure the latency level. These servers are called metropolitan edge

M. S. A. Muthanna (✉) · D. Elkin · S. Likhtin
Institute of Computer Technologies and Information Security, Southern Federal University, Taganrog, Russian Federation
e-mail: muthanna@sfedu.ru; delkin@sfedu.ru

A. M. Al-Sveiti Malik
The Bonch-Bruevich Saint-Petersburg State University of Telecommunications, St. Petersburg, Russian Federation

A. A. Abd El-Latif et al. (eds.), *Secure Edge and Fog Computing Enabled AI for IoT and Smart Cities*, EAI/Springer Innovations in Communication and Computing,
https://doi.org/10.1007/978-3-031-51097-7_1

computing (MEC) servers [5, 6]. This chapter will compare a single-level and a two-level MEC system.

Autonomous vehicles have the potential to greatly improve safety and traffic flow, in the future, when autonomously driven vehicles become ubiquitous. It is important to note that accomplishing this goal will necessitate major investment in machine learning models for traffic interpretation as well as infrastructure for low-latency wireless networks. While 5G mobile networks are still in the planning stages, it will still be a few years before unmanned cars are connected to 5G. For now, carriers are investing in 5G networks, and thus dependent autonomous driving research is accelerating. With autonomous vehicles, there are numerous approaches—or potential results. V2V communication, as the name implies, connects vehicles to each other to exchange route and speed data. V2X connectivity provides greater connectivity, allowing automated communication with parking meters, parking garages, or for example, if there are no vehicles traveling on a particular road, there is a possibility that the street lights get switched off. In order to build these connections, not only must the technology be integrated into the vehicles themselves, but there must also be a significant rethinking of the way that public infrastructure is constructed.

For the deployment of a 5G-enabled autonomous vehicle, the widespread availability of 5G networks is required. Tesla is currently delivering semiautonomous driving capabilities, although networks are used only as a supplementary form of communication [1]. For all road systems to work properly and for unmanned cars to understand their place in the environment, 5G networks must be ubiquitous.

In order to start using autonomous car technology, many experts agree on the need for large-scale adoption of 5G as the successor of wireless technology. The present 4G network is fast enough for HD video streaming and online gaming, but it cannot enable safer and more intelligent autonomous vehicles. The current 4G network is not fast enough to provide instantaneous operation of new services for data processing by autonomous vehicles at the level of human reflexes.

2 V2X Technology Description

V2X is an umbrella word for a vehicle communication system that transmits data from sensors and other sources through channels with high bandwidth, low latency, and high dependability, opening the way for completely autonomous driving. V2X includes V2V, V2I, V2P, and vehicle-to-network (V2N) communication. In this multidimensional ecosystem, cars will communicate with other cars, with infrastructure such as traffic lights or parking lots, with pedestrians using a smartphone, and with data centers via cellular networks. It is necessary for the communication system to be both efficient and cost-effective to manage varying sets of requirements based on various purposes. Figure 1 displays the various V2X communication methods.

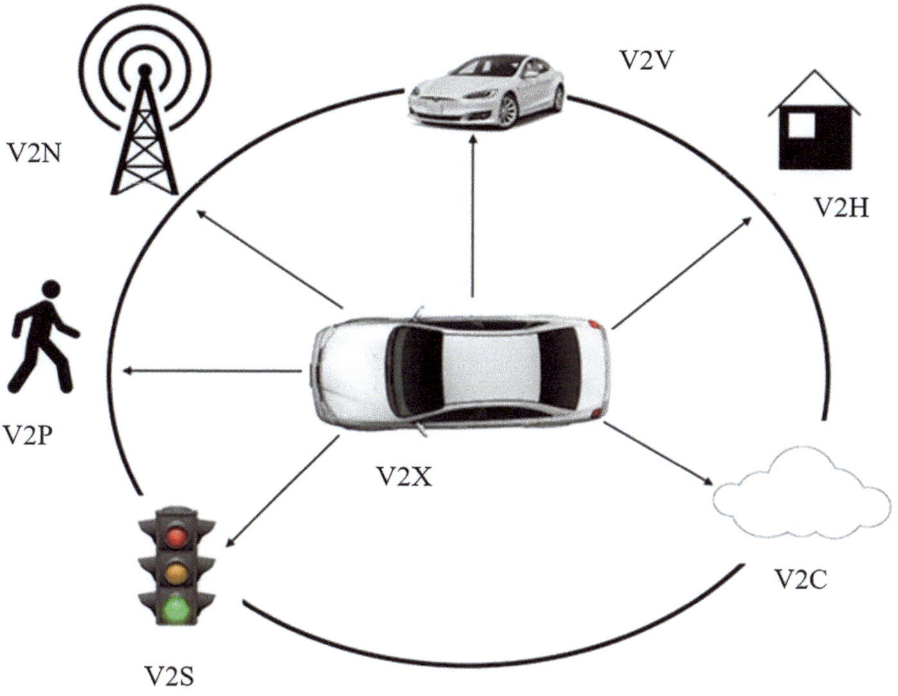

Fig. 1 V2X types technology

Connected vehicle technologies and autonomous vehicle technologies are frequently viewed as different technologies, yet they are actually orthogonal. Good communication between these two technologies improves their interaction for the intelligent transportation system's reality. Many applications can be supported by V2V and V2I communication. Collaboration between vehicles and infrastructure can be extended to applications, for example, to organize a joint group, where vehicles share their mobility data to maintain group integrity, or for a joint traffic information system, where vehicles share their scheduled beacons to build traffic projections. Similarly, autonomous cars can also communicate with each other in other situations. For example, sharing sensor data with other road users would not only help them in various situations, such as maneuvers, but also make driving safer. Cooperation between autonomous vehicles can have many other important benefits. For example, cooperative state estimation with the exchange of trajectory information between autonomous vehicles can make navigation more efficient and smoother, as well as help autonomous vehicles in case one of its subsystems fails. Cooperative localization through improved sensor design and motion coordination are other applications of cooperative communication between autonomous vehicles.

The original V2X protocol is based on a Wi-Fi offshoot, IEEE 802.11p (part of the IEEE WAVE or wireless access for automotive environment initiative), which operates in the unlicensed 5.9 GHz frequency band. The 2012-completed IEEE 802.11p standard supports dedicated short-range communications (DSRC) in the United States and ITS-G5 as part of the European Cooperative Intelligent Transportation Systems (CITS) effort. V2X communication using 802.11p extends beyond line-of-sight constrained sensors like cameras, radar, and Light Detection and Ranging (LIDAR) to cover V2V and V2I use cases such as collision warnings, speed restriction alerts, and electronic parking and tolls. These use cases span a wider range of applications. 802.11p is said to have a short range (less than 1 km), low latency (2 ms), and high dependability, as stated by the United States Department of Transportation; it "operates in high vehicle speed situations and provides resistance to adverse weather conditions (e.g., rain, fog, snow, etc.)." 802.11p boosts a vehicle's ability to "see" its surroundings, especially in adverse weather conditions. IEEE 802.11p is independent of cellular network coverage and solutions for onboard unit (OBU) and roadside unit (RSU).

C-V2X or cellular V2X is a promising alternative to IEEE 802.11p, supported by the 5G automobile association and chip manufacturer Qualcomm. C-primary V2X's advantage is its two modes of operation, which cover the vast majority of circumstances. The first is the C-direct, V2X's low-latency PC5 interface communication in the unlicensed 5.9 GHz band, which is designed for active safety messaging like immediate traffic warnings and other near-field V2V, V2I, and V2P circumstances. This mode is quite similar to the existing IEEE 802.11p technology, which likewise utilizes the 5.9 GHz frequency spectrum. The second mode, Uu interface communication over a conventional cellular network with a licensed band, is capable of handling V2N use cases such as information entertainment and delay-tolerant safety alerts regarding longer-range road hazards or traffic conditions. IEEE 802.11p can only comply with this mode by establishing unique links to roadside base stations because it does not utilize cellular communications. 5GAA, the consortium supporting C-V2X, debuted in 2017 with eight founding members. There are presently 120 members. Audi, BMW, Daimler, Ford Motor Company, General Motors, Honda, Hyundai, Nissan, VW, and Volvo are some of the companies that make up the membership. Similarly, companies in the technology industry such as Intel, Samsung, and Qualcomm, companies that manufacture auto electronics such as Alpine, Continental, and Bosch, companies that manufacture network equipment such as Nokia and Ericsson, and wireless carriers such as AT&T, T-Mobile, Verizon, and Vodafone are all examples of multinational corporations. C-V2X is currently compatible with 4G networks. 3GPP, an industry consortium that sets standards for wireless networks, has announced C-V2X technology that enables vehicles to transmit basic driving data over 4G networks. 6G networks will have ultrahigh density, ultralow latency, and high data rates.

The main characteristics for 5G are frequency aggregation, multi-RAT radio access technology, power consumption, signal latency, reliability and security, and SD-RAN.

Categories of services in 5G that are considered essential:

- Ultra-broadband mobile connectivity, which provides high bandwidth and data rates.
- Ultrareliable, low-latency machine-to-machine communication—this type of service stands out for its low data latency (<1 ms per side) with very high connection reliability and availability.
- Massive machine-to-machine communication is in the field of application services and applications with the ability to connect a very large number of Internet of Things (IoT) devices.

V2X technologies were considered. Vehicle-to-vehicle (V2X) communications system includes some types of communications such as:

- vehicle-to-vehicle (V2V),
- vehicle-to-infrastructure (V2I),
- vehicle-to-pedestrian (V2P),
- vehicle-to-network (V2N).

The two standards on which V2X communications are based were reviewed. The 802.11p standard extends beyond line-of-sight-restricted sensors such as cameras, radar, and LIDAR to include V2V and V2I use cases. Also mentioned were the benefits of the C-V2X or cellular V2X standard over 802.11p. C-primary V2X's advantage is that its two modes of operation cover the vast majority of probable situations. By using 4G or 5G mobile cellular, the C-V2X has a wider range of capabilities compared to DSRC 802.11p. The adoption of C-V2X technology by various automotive manufacturers has also been analyzed.

3 The Advantage of Edge Computing Technology

Today, edge computing defines the network's future. They offer devices with stability and alleviate latency issues by processing data closer to the source.

MEC defined edge computing as a network of small data centers that store and process mission-critical data on the premises. Edge computing is also known as fog computing. These data centers are connected as a network, and they deliver the data that they receive to a centralized data repository. In most cases, this takes place within a ten-square-meter area.

IoT edge computing is utilized to analyze and process data in a location that is geographically closer to its source. IoT edge computing employs intelligent gadgets that are able to process vital data and give prompt, real-time solutions. These devices eliminate the delays created by transferring data to the cloud over the Internet and by the cloud's delayed response.

These devices are intended to function as miniature data centers with near-zero latency. This enhanced capacity decentralizes data processing and significantly reduces network traffic. The cloud can then collect these data for additional analysis and processing.

Types of IoT edge computing.

In general, three types of IoT edge computing can be distinguished:

- Devices that are located locally to serve a definite and well-articulated purpose. They are easy to deploy and maintain.
- Data centers located locally to provide a significant amount of processing and storage capacity. They are typically pre-engineered and preconfigured. They are assembled in a specific location and provide good economies of scale.
- The fact that regional data centers are situated in close proximity to the actual sources of data is a significant benefit enjoyed by these facilities. They are more expensive and require more upkeep than local data centers, but they offer greater processing and storage capacity than those facilities.

Large numbers of sectors are realizing their potential benefits as IoT edge computing is embraced and becomes popular. Specifically, edge computing offers seven significant benefits for smart manufacturing:

1. Quick reaction times: Local and distributed compute and storage power. Avoiding transferring data back and forth to the cloud is essential for decreasing latency and improving reaction times. This will assist in preventing harmful events.
2. Operating with unreliable connectivity: Remote agricultural pumps, oil wells, windmills, and solar farms, it can be difficult to govern variable areas of Internet connectivity. Edge's local storage and processing capabilities prevent data loss and outages in the event of intermittent Internet connectivity.
3. Security: IoT edge computing eliminates cloud device data transfer. Filter sensitive data locally and transmit the data model to the cloud.
4. Cost-effective: The main practical challenge in implementing the Internet of Things has been the costs associated with data storage, computing power, and network bandwidth. IoT Edge computing lets businesses distinguish between local and cloud-based offerings by calculating data locally. This lowers IoT solution development costs.
5. Edge devices operate as a bridge between current and legacy equipment to ensure compatibility between the two. This enables legacy equipment to communicate with contemporary devices for IoT solutions.
6. Decentralized edge computing makes other network devices more resilient. This is important since a single cloud server failure can damage thousands of IoT devices.
7. Reduced data exposure: Edge computing decreases network-wide data transmission. This decreases transit data exposure. PII and payment card industry data may not be transferred. Such situations will aid in avoiding certain security and privacy-related legal issues. In addition, data encryption and access controls can bolster its protection against common threats.

4 Model Description

Mobile edge computing (MEC) allows us to perform operational calculations at a close range of terminal devices, which can also be unmanned vehicles [7, 8]. This technology significantly reduces latency and unloads the network, making it possible to use technologies such as the IoT, augmented reality, and holography.

In our model, three levels will be used. The first level includes terminal devices, that is, cars equipped with onboard units (OBU). The second layer consists of edge computing servers at the edge of the radio access network. This level consists of two types of servers: a MicroCloud server close to the endpoints and a mini-cloud server that combines several MicroClouds. The third layer is a central cloud server located far from the terminal devices but has a large computing area. This structure is shown in Fig. 2.

Next, let's take a closer look at the second level of our system. As mentioned earlier, it consists of two levels. The first layer is micro-cloud servers that receive a signal from a base station that receives signals from target devices, and this station forms a small area that this server will handle. Several micro-cloud servers, in turn, are connected to the mini-cloud Server using an antenna that forms beams and serves micro-cloud servers.

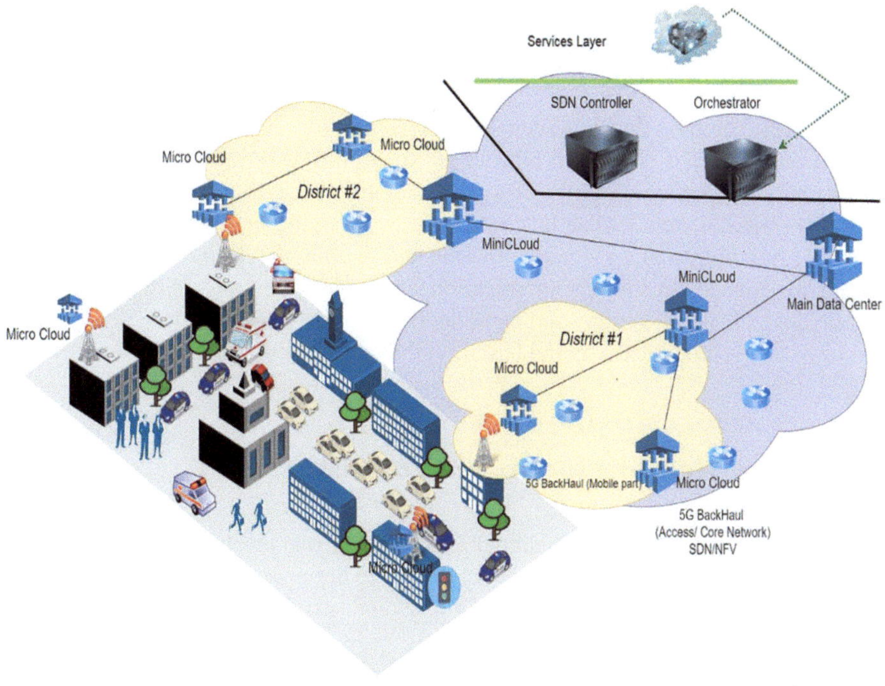

Fig. 2 Multilevel cloud system

Mini-cloud servers reduce the delays of data packet transmission. It is proposed, like micro-cloud servers, to be placed close to the border of the radio access network. The micro-cloud can be deployed near the base station, through which data streams from user devices to this micro-cloud are transmitted or in the coverage area of the base station. The mini-cloud, in turn, can be deployed near the base station controller. This base station controller recommends pairing a mini-cloud with micro-clouds that receive signals from base stations. A mini-cloud can also be organized at the logical level of a base station controller but without connecting to a specific controller. In this case, the mini-cloud will be used to combine micro-clouds receiving signals from the base stations of several different base station controllers.

If, for example, micro-cloud servers do not cope with the flow of data in their area, they send this data to the next level. Therefore, processing the data flow will be a system of queues. Requests that have not been processed promptly at one level will be carried over to the next level until they reach the highest level.

The third level of the proposed system, along with the central cloud service, is proposed to place the Software-defined networking (SDN) controller and orchestrator.

Management of logical data transfer from micro-cloud to mini-cloud and from mini-cloud to central cloud server is offered to be carried out using the controller of software-configurable SDN networks [5]. SDN data flow management will also require the implementation of SDN switch functions on micro-clouds and mini-clouds.

Such a system can be used both for the VANET network, in which all requests are of single priority, and for fog computing, in which requests can be of different priorities.

It is worth noting that the second-level computing clouds of the architecture offered by us (micro-cloud and mini-cloud) can be transferred to the functions of the fog server, forming a cluster of fog devices for processing data streams [8]. At the same time, fog devices can be considered smart traffic lights, video surveillance cameras, and information stand implemented within the framework of another promising concept in info-communication networks—the smart city concept and mobile devices of pedestrians and passengers of public vehicles [9].

For example, consider a model based on the VANET environment.

5 Analyze Modeling for the Environment VANET

A model was created that shows signal processing at two levels of edge computing. Second-level requests are those requests that were not processed at the first level.

The distribution of response for single-level and two-level systems obtained from the simulation results can be seen in Fig. 3.

Traffic going to the second level depending on the time is shown in Fig. 4.

The simulation results clearly show that the two-level system is more stable; with the help of its use, it is possible to significantly increase the processing speed and

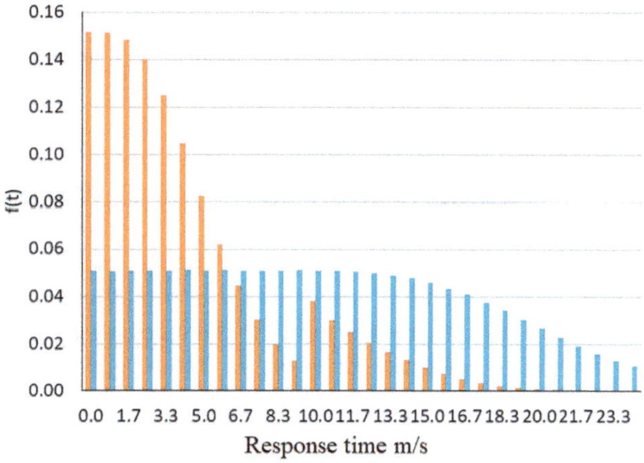

Fig. 3 Distribution of response time in a two-tier system

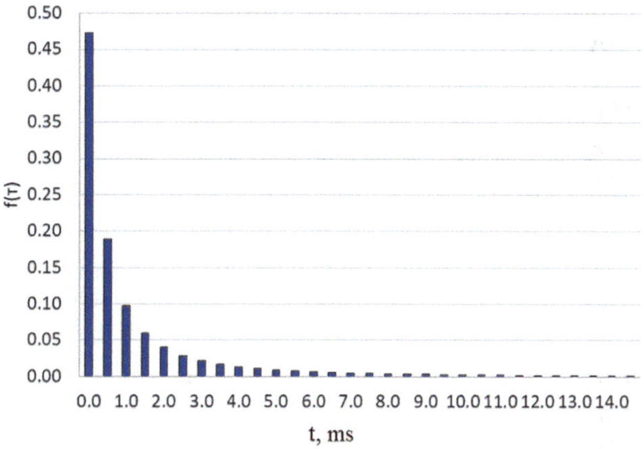

Fig. 4 Distribution of the time of arrival of the traffic on the second level

significantly reduce the latency in the network, which was the purpose of developing such an architecture.

6 Conclusion

Edge computing technology (MEC) can be an excellent solution for networks such as VANET. This will improve the characteristics of this local network served by the edge servers. But sometimes, such a system cannot cope with the load, which is constantly changing, due to the high mobility of users. For such a problem, a multilevel system of edge computing is perfect. The layer server will handle unnecessary traffic to offload the network and prevent the lower-level servers from the lower level. In this chapter, a model was created that processes the signals of these levels. The simulation results showed that a two-level system could raise the processing speed and reduce the latency in the network.

It is also worth noting that the proposed system also has high integration capabilities into the concept of a smart city, as it will allow the processing of large amounts of data from video surveillance systems, information stands, and smart traffic lights while ensuring low values of delays in the transmission of data packets. The proposed system will also allow more efficient organization of fog clusters since fog servers (compared to the data center) have a closer location to fog devices, which will allow timely information about their free resources (RAM, communication channel bandwidth, battery charge, CPU).

The proposed architecture in the future may contribute to the introduction of various services for passengers of ground vehicles, as it will allow processing data coming from mobile devices of users with high passenger traffic.

Acknowledgments The research is supported by postdoc fellowship granted by the Institute of Computer Technologies and Information Security, Southern Federal University, project № P.D./22-01-KT.

References

1. Muthanna, M.S.A., Wang, P., Wei, M., Rafiq, A., Josbert, N.N.: Clustering optimization of LoRa networks for perturbed ultra-dense IoT networks. Information. **12**, 76 (2021)
2. Muthanna, M.S.A., Muthanna, A., Rafiq, A., Hammoudeh, M., Alkanhel, R., Lynch, S., Abd El-Latif, A.A.: Deep reinforcement learning based transmission policy enforcement and multi-hop routing in QoS aware LoRa IoT networks. Comput. Commun. **183**, 33–50 (2021)
3. Muthanna, M.S.A., Wang, P., Wei, M., Abuarqoub, A., Alzu'bi, A., Gull, H.: Cognitive control models of multiple access IoT networks using LoRa technology. Cogn. Syst. Res. **65**, 62–73 (2021) ISSN 1389-0417
4. Chaaf, A., Saleh Ali Muthanna, M., Muthanna, A., Alhelaly, S., Elgendy, I.A., Iliyasu, A.M., et al.: Energy-efficient relay-based void hole prevention and repair in clustered multi-AUV underwater wireless sensor network. Secur. Commun. Netw. **2021**, 1–20 (2021)
5. Rafiq, A., Ping, W., Min, W., Muthanna, M.S.A.: Fog assisted 6TiSCH tri-layer network architecture for adaptive scheduling and energy-efficient offloading using rank-based Q-learning in smart industries. IEEE Sensors J. **21**(22), 25489–25507 (2021)
6. Muthanna, M.S.A., Alkanhel, R., Muthanna, A., Rafiq, A., Abdullah, W.A.M.: Towards SDN-enabled, intelligent intrusion detection system for Internet of Things (IoT). IEEE Access. **10**, 22756–22768 (2022). https://doi.org/10.1109/ACCESS.2022.3153716
7. Ahsan, R., Ping, W., Min, W., Ali, M.M.S., Nkerabahizi, J.N.: Mitigation impact of energy and time delay for computation offloading in an industrial IoT environment using Levenshtein distance algorithm. Secur. Commun. Netw., 1939–0114 (2022)
8. Aboulola, O., Khayyat, M., Al-Harbi, B., Muthanna, M.S.A., Muthanna, A., Fasihuddin, H., Alsulami, M.H.: Multimodal feature-assisted continuous driver behavior analysis and solving for edge-enabled internet of connected vehicles using deep learning. Appl. Sci. **11**, 10462 (2021)
9. Chiu, S.N., Stoyan, D., Kendall, W.S., Mecke, J.: Stochastic Geometry and Its Applications, 583p. Wiley (2013)

UAV-Based Edge Computing System for Smart City Applications

Mehdhar S. A. M. Al-gaashani, Alexander Speransky, Muthana Ali Salem, and Alexey Tselykh

1 Introduction

In the current era marked by the development of 5G networks, a multitude of projects and ideas, previously unattainable with earlier mobile network generations, are now coming to fruition. The concepts of smart home, smart city, smart enterprise, and tactile internetworking had been a part of the research, but it was difficult to implement. The revolution accompanied by 5G technologies has enabled the realization of three services [1–5]:

- Creation of ultra-broadband mobile communications (EMBB);
- Creation of ultrareliable communication with low latency (URLLC);
- Support for mass machine-to-machine communication (mMTC) [6].

M. S. A. M. Al-gaashani (✉)
College of Computer Science and Technology, Chongqing University of Posts and Telecommunications, Chongqing, China

A. Speransky
The Bonch-Bruevich Saint-Petersburg State University of Telecommunications, St. Petersburg, Russian Federation

M. A. Salem
National University of Sciences &Technology Leninskiy Prospekt 4, NUST MISIS, Moscow, Russia
e-mail: m2112648@edu.misis.ru

A. Tselykh
Institute of Computer Technologies and Information Security, Southern Federal University, Taganrog, Russian Federation
e-mail: tselykh@sfedu.ru

A. A. Abd El-Latif et al. (eds.), *Secure Edge and Fog Computing Enabled AI for IoT and Smart Cities*, EAI/Springer Innovations in Communication and Computing,
https://doi.org/10.1007/978-3-031-51097-7_2

These services have become the main catalyst for the smart objects' further research.

The global goal of 5G services is to develop the telecommunications industry for quality improvement and the management of society's life processes simplification. This goal is achieved at the level of houses, at the level of enterprises, and at the level of cities [7].

The smart city is an idea that improves many processes in the urban environment: optimization of the transport system operation, energy efficiency, and saving resources and enhances the safety of citizens. A smart city is a city that aims to achieve the goals of the city of the future through the use of information and communication technology (ICT) solutions and trends. The implementation of 5G based on unmanned aerial vehicles' use can solve these tasks. The next chapter explores the role of UAVs in smart city design and analyzes the network architecture model that supports UAV applications in conjunction with 5G services [8–10].

2 Network Architecture Using MEC/SDN Technologies and UAVs as Edge Nodes

UAVs were previously used only for military purposes, but they currently have become widely used in civilian industries. UAV implementation in smart cities is a promising area of application.

It is shown in McKinsey and Company report that global spending on urban construction and infrastructure is about 2 trillion US dollars per year, and ICT spending is between 1.5% and 2% of this number [11]. It is expected that over the next decade, advances in cloud computing, wireless sensors, networked drones, big data, open data, and the Internet of Things will continue to develop. Therefore, it is expected that it would be possible to use UAVs to solve cities' problems.

It is planned that using UAVs will perform the following functions in smart city tasks:

- Traffic management. Airborne drones can assess traffic conditions in cities and relay the information to telecom hubs for use in traffic management applications. This feature will help almost completely to get rid of traffic management problems.
- Gathering information from sensory nodes. Using UAVs as a mobile gateway will allow monitoring of the objects of urban infrastructure hazardous to human health and life.
- Disaster management and situation monitoring. Possibility of a high-resolution shooting near objects from low altitudes for the use of UAVs in emergency areas without any risk to human life and health.
- Ensuring public safety. With the ability to transmit photographic images in real time, UAVs could be used to pursue targets, conduct rescue operations, etc.
- Transportation of goods. UAVs may become one of the ways to deliver small-sized cargo, correspondence, etc.

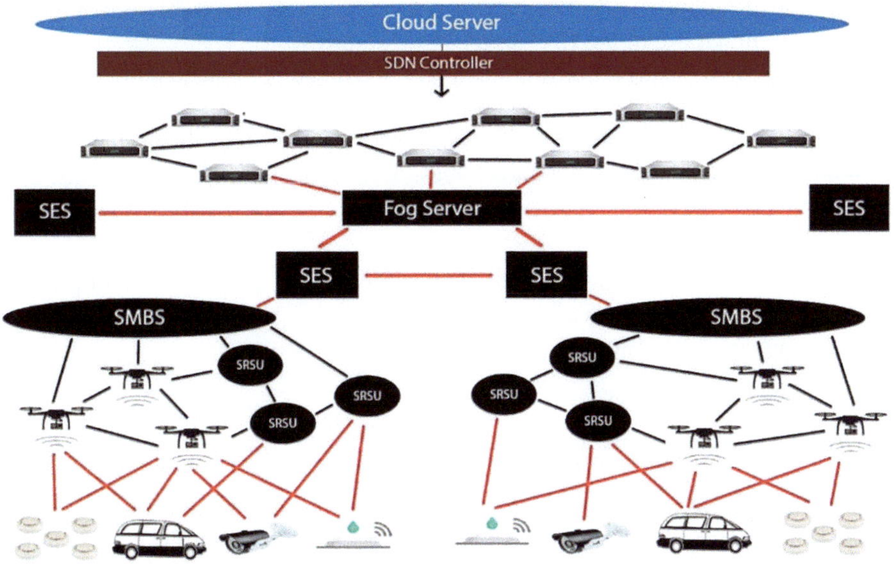

Fig. 1 UAV-based smart city network architecture

For the effective UAVs used for smart city applications, it is necessary to create a network that will provide reliable, low-latency communications for a huge number of devices. It is necessary to build an architecture based on local computing to offload traffic from the network core because of the huge amount of traffic that UAVs must process. This will help make calculations of local tasks faster and offload the network core traffic. It is necessary to deploy SDN technology along the entire traffic path to optimize the traffic path for cloud data [2].

Based on the above considerations, an architecture of the optimal UAVs used for the smart city tasks was proposed (Fig. 1).

Smart city tasks require fast and ultralow latency computing, such as traffic management. These tasks cannot be solved by computing all the data in cloud centers. Despite the huge computing power of the data center, there are large restrictions on the bandwidth of networks with traditional architecture. This problem becomes more relevant in the Internet of Things development era. That is why it was decided to use mobile edge computing (MEC) technology for the smart city network architecture. MEC's task is to calculate the data at the network edge, and it is proposed to process data that does not require real-time calculations in the cloud. MEC technology organizes three levels of computing in the proposed architecture: cloud server, fog server, and edge server.

As opposed to MEC, SDN provides efficient load distribution between network nodes due to the separation of control functions and physical data transfer. SDN is used in a hybrid model that is implemented in the network core and in edge computing servers (fog server, Satellite Earth Station (SES), Small and Medium-sized Businesses (SMBS)). When implementing SDN in SMBS and SES, the problem of geographic optimization of the UAV is solved for traffic offloading or for the operational use of the UAV as a mobile gateway.

Thus, based on the proposed architecture, it would be possible to build a network capable of solving all the smart city tasks.

Despite the benefits of using UAVs, there are a number of problems that prevent its implementation. UAVs' problems could be divided into two areas – legal and technical.

Legal issues include:

1. Confidentiality: UAVs are considered by many experts as an invasion of privacy.
2. Price: Deploying UAVs is an expensive task.
3. Licensing and legislation.

Technical issues include:

1. UAV middleware: Software that is necessary for the uninterrupted and correct operation of the UAV.
2. Development of fault-tolerant systems: Systems must guarantee a high level of safety in aircraft failure and loss of communication between the UAV and the control center.
3. Accurate UAV flight in terms of altitude and flight path over long periods of time in all weather conditions, both day and night.
4. Development of UAV perception mechanisms: These mechanisms are necessary for the perception of the environment, allowing the UAV to take evasive actions [3].

3 Development of a Network Model with Edge and Cloud Computing Nodes

The system consists of two levels of calculations:

1. Edge computing: In this model, the layer is represented as a local machine running in a docker container. The local machine is two times more resource constrained than a Linux cloud virtual machine (VM).
2. Cloud computing: In this model, the level is a virtual machine running in the cloud platform. A cloud platform is a virtual machine deployed in two physically possible regions where an application instance is located: region eu-west-2 (London) and sa-north-1 (Brazil). Placing the application in the sa-north-1 region simulates running the application on a high-load network with high network latency anomalies.

This experiment runs a high-load application in both layers and compares the use of the edge and cloud layers for low-load and high-load smart city networks.

In order to model the selected network, it was decided to use the tools that are actively used for web application development. In order to simulate the application operation, it was decided to use Java, which is used to write application logic that can load the processor and RAM of the computing node.

To interact with the application via the REST (Representational State Transfer) API (Application Programming Interface), it was decided to use Spring Boot framework, which implements the function of processing requests and sending server responses. The framework uses the built-in Tomcat server as a web server.

Docker was used to isolate the application from the hardware, and it allows to abstract the launch of the application from the physical machine.

Amazon Web Services (AWS), used as a cloud platform, provides opportunities for deploying remote applications (selection of the region, resources of the instance being launched).

As a tool for generating requests and calculating delays between requests and responses, it was decided to use the JMeter tool to flexibly configure various parameters of Rest requests (number of virtual users, time between requests, number of iterations between requests).

A model based on the above technologies was chosen for the experiment.

The main task of the application is to load the maximum resources of the computer. An application that cyclically processes large images was written for these purposes. Figure 2 shows an application that writes the data obtained during image processing into the machine's local memory. This application is used both on the local computer and on the cloud. During program execution, threads are created within the application. In this case, there is an attempt to run 100 threads (an attempt to run on 100 cores), leading to a "deficit" of processor computing time for each thread. In addition to creating many threads, the application also uses a large number of try-catch statements, which are "heavy" for executing a Java program. In addition to the above two methods of loading computer resources, the process of writing decoded bytes to a file in the machine's memory is also launched on each thread.

Spring boot in this experiment performs the most important function of auto-configuring the server-side of the application (embedded web server configuration, opening a port for accepting requests, redirecting requests to Java classes (controllers)). Figure 3 shows the application logic written on top of the Spring boot framework that is responsible for accepting requests from the client.

The @RestController annotation marks the controller class as the class responsible for accepting requests on the server side. This annotation also informs that the response to the client will be generated in JSON format.

The @RequestMapping annotation "tells" the framework the address to redirect it to the server. In this case, for the server to accept the request, it is important to specify API addresses in the URL. If the application is run on a local machine, the server will be launched at localhost:8080 by default. If it is necessary to process the request locally, localhost:8080/api must be contacted. After accepting the request, the server then redirects to specific methods, which may also have an addition to the address in the URL. The methods handle each type of request in the REST API (POST, PUT, GET, DELETE). In the developed application, there are two types of POST and GET requests which are marked with @PostMapping and @GetMapping annotations, respectively.

```
public String processfile(MultipartFile string){
    Thread thread = new Thread(new Runnable() {
        @Override
        public void run() {
            for(int i = 0; i<100; i++) {
                byte [] fileBytes  = new byte[0];
                try {
                    fileBytes = string.getBytes();
                } catch (IOException e) {
                    e.printStackTrace();
                }
                FileOutputStream outputStream = null;
                try { outputStream = new FileOutputStream( name: "file.txt");
                } catch (FileNotFoundException e) {
                    e.printStackTrace();
                } try { outputStream.write(fileBytes);
                } catch (IOException e) {
                    e.printStackTrace();
                } try { outputStream.close();
                } catch (IOException e) {
                    e.printStackTrace();
                }
            }
        }
    });
    thread.start();
    return "Task is finished";
```

Fig. 2 Program logic reading image data and writing to machine memory

The method that accepts a GET request puts an additional/test address in the url. This method is used to test the receipt of a request from the client and the correct response of the program to requests from the outside. This method returns the server response in the form of the text "It's ok."

The method accepting the POST request is responsible for accepting the MultipartFile (picture) from the client side. The passed MultipartFile parameter is required in requesting. POST was chosen as the request type since it is possible to encapsulate the image data in the body of the HTTP request method. After accepting the request, the method passes data to the service layer to the process file() method (the content of this method is described earlier). After successful data processing, the method returns "Task is finished," which is passed to the client to respond to a request to the server. The server returns an HTTP response with a processing status of 200, indicating that the server correctly responded to the user's request.

```
@RestController
@RequestMapping("/api")
public class Controller {
    private Service service;
    public Controller(Service service) { this.service = service; }

    @PostMapping
    public String getResponse(@RequestParam (value = "string") MultipartFile string)
            throws InterruptedException, IOException {
        return service.processfile(string);
    }

    @GetMapping("/test")
    public String getResponseTest() {
        return service.test();
    }
}
```

Fig. 3 Application logic is responsible for accepting requests from the server

Docker is the most important tool in the experiment, which provides complete isolation of the developed system from the running hardware and also flexibility when deciding to port the application to third-party physical servers.

After writing the server logic and the application's business logic, it is necessary to isolate the program from the physical environment and place the application in the image. The application can be hosted and transferred over the network as a file containing all the necessary information about the application and its dependencies.

We use the command docker build, which packs the entire environment into an image and assigns a port that binds the port of the docker container and the port of the application itself inside the docker container. The application will be available on port 8080 of the docker container in this experiment.

The application compiled into the image must be available for installation on cloud servers. For this, we use elastic container repository (ECR), which stores the image in a dedicated repository along with tags (provides image versioning). As part of the experiment, the image is pushed to the repository in ECR and tagged "latest" using the command docker push <repository name>:latest command. After these steps, the image is ready to be deployed to the server.

In the developed network model, Amazon Web Services are cloud infrastructure. This network must fulfill the requirement to deploy the application in the cloud infrastructure. AWS has the capability to host an application in a remote repository (ECR) and directly provide physical resources for deploying an application on a virtual machine.

The previous paragraph described the process of hosting a docker image in the ECR repository. Once the built application image is uploaded to the repository, it is available for deployment to a physical EC2 server. To successfully deploy and run the application on a cloud server, it is necessary to describe the AWS infrastructure.

```
const vpc = new Vpc(this, "VPC-UI", {
  vpcName: 'VpcUI',
  subnetConfiguration: Vpc.DEFAULT_SUBNETS_NO_NAT,
  maxAzs: 3
})

const cluster = new Cluster(this, 'shopping-ui-cluster', {
  vpc: vpc,
  capacity: {
    instanceType: InstanceType.of(InstanceClass.BURSTABLE3, InstanceSize.MICRO),
    desiredCapacity: 1,
    associatePublicIpAddress: true,
    vpcSubnets: {
      subnetType: SubnetType.PUBLIC
    }
  }
})

new ApplicationLoadBalancedEc2Service(this, 'Ec2ShoppingUI', {
  cluster: cluster,
  healthCheckGracePeriod: Duration.hours(4),
  cpu: 512,
  memoryLimitMiB: 700,
  publicLoadBalancer: true,
  taskImageOptions: {
    environment: {
      REST_API_URL : 'localhost:8081'
    },
    image: ContainerImage.fromEcrRepository(
        Repository.fromRepositoryName(this, 'Repository', 'diploma-application'),
        '9e71c204720c8bcadf40371c8fc988f22f9e048fsa-east-1'
    ),
    containerPort: 8081
```

Fig. 4 Code describing the Amazon Web Services infrastructure

The AWS CDK tool has been used to describe the AWS infrastructure. Figure 4 shows the code that describes the AWS infrastructure that will interact with AWS services through the AWS CDK.

Figure 4 describes three main services in AWS:

1. VPC: Amazon Virtual Private Cloud (VPC) allows you to run AWS resources in a specific virtual network. This virtual network is very similar to the traditional network you would use in your own data center, with the benefits of using the scalable AWS infrastructure. In this network model, we chose the Vpc.DEFAULT_SUBNETS_NO_NAT configuration. This configuration creates one public and one isolated subnet in each accessibility zone. A maximum of three availability zones are created in the created virtual network.

2. Cluster: An Amazon ECS cluster is a logical grouping of tasks or services. Tasks and services run on the infrastructure registered with the cluster. Infrastructure capacity can be provided by AWS Fargate, a serverless infrastructure that manages AWS Amazon EC2 instances managed by either a local server or a virtual machine (VM). In the developed network model, it was decided to use the EC2 cluster type, so it is necessary to predetermine the maximum allowable size of EC2.

   Cluster settings:

- Services predefined in this cluster are launched in a previously configured VPC (point 1).
- InstanceType is chosen as T3.micro. This instance type provides a base layer of performance. T3.micro has the following specifications: 2 vCPUs, 12 CPU credits per hour, 1 GB memory, Elastic Block Store (EBS)-only storage, performance up to 5 Gbps, Intel Xeon Platinum processor (up to 3.1 GHz); there is an optimization for EBS, improved network configuration. The least productive instance to be able to load the CPU and RAM as much as possible is chosen for the developed network model.
- DesiredCapacity is 1. The maximum number of instances of running virtual machines cannot exceed 1. This setting is used to configure AutoScaling in order to increase computing power as the load increases. In the developed network model, this function contradicts the conditions of the experiment.
- AssociatePublicIpAddress determines whether the instances configured in this cluster have a public IP address. A public address is needed to access it from external networks in this network model.
- VpcSubnets determines which subnet the virtual machine instances will run on. As described earlier, one public and one isolated subnet were created within the created VPC. It is on the public subnet that the configured instances will run.

3. ApplicationLoadBalancedEc2Service: As part of this AWS resource, a virtual machine instance is directly created on which the developed application is launched. This resource describes an EC2 service running on an ECS cluster with an application load balancer. Although the EC2 instance itself has a public IP address, all traffic arriving on the EC2 instance will be intercepted by the load balancer, which, in turn, directs incoming requests to EC2. The load balancer automatically distributes incoming traffic to multiple targets such as EC2 instances, containers, and IP addresses in one or more availability zones. It monitors the health of its registered targets (instances) and directs traffic only to healthy targets. Elastic load balancing supports the following load balancers: application load balancers and network load balancers. The model will use an application load balancer. The EC2 load balancer has the following settings within the network model:

   - The cluster in which the EC2 instance is running. This cluster has been described previously.

- HealthCheckGracePeriod is set to 4 h. It defines the period of time in which the status of the service is checked. If the service is not available after the expiration of this time, then an attempt is made to start the described service again.
- CPU – the number of processor units used by the task being launched.
- MemoryLimitMib defines the maximum RAM allocated to running containers within a running instance. In the running model, the memory limit is set to 700.
- PublicLoadBalancer determines the availability of the load balancer from external services and the allocation of a Domain Name System (DNS) address for it.
- TaskImageOptions is one of the most important parts in setting up an EC2 instance, defining the image of the application image uploaded to the ECR docker service. As part of the "environment" setting, global variables are passed that are used inside the docker application. The "image" setting describes the docker image of the application that needs to be deployed as part of the running instance. This specifies the name of the repository in ECR, the name of the image within the repository, and a tag that describes the version of the application image. It also describes containerPort as 8081. The application described earlier in taskImageOptions is launched on port 8081.

The AWS Cloud Development Kit (CDK) deploy command successfully deploys the already-built application to the Amazon server. Access to the application is possible by public DNS address. To deploy the application, you also need to configure the hosting region. As part of the network model being developed, the application is deployed in two regions: eu-north-1 (Stockholm) and sa-north-1 (Sao Paulo).

Jmeter is a tool for functional and load testing. JMeter has the necessary supporting protocols for testing Rest applications: Hypertext Transfer Protocol (HTTP), Simple Object Access Protocol (SOAP), File Transfer Protocol (FTP), Lightweight Directory Access Protocol (LDAP), Simple Mail Transfer Protocol (SMTP), Post Office Protocol version 3 (POP3), and Internet Message Access Protocol (IMAP). JMeter also has the functionality of queuing requests in the Thread Group settings. The Thread Group has the ability to set the number of iterations in the request configuration. The behavior of a thread is determined according to the scaling and is destroyed after a certain number of iterations per thread. Under the list of executed queries, there is a variety of visualization of test results in the form of diagrams, tables, etc.

Within the framework of the executed model, the following settings are set in the Thread Group: 1 thread, 1 s – time interval between requests, 50 iterations, 1 – the number of users executing the request (Fig. 5).

After creating a request flow configuration, which is a common setting for both on-premises and cloud applications, you need to create separate request configurations for both on-premises and cloud applications.

The local application is run on the local machine using the docker environment on port 8081. Docker assigns the IP address as 172.17.0.2. Using it, jmeter will

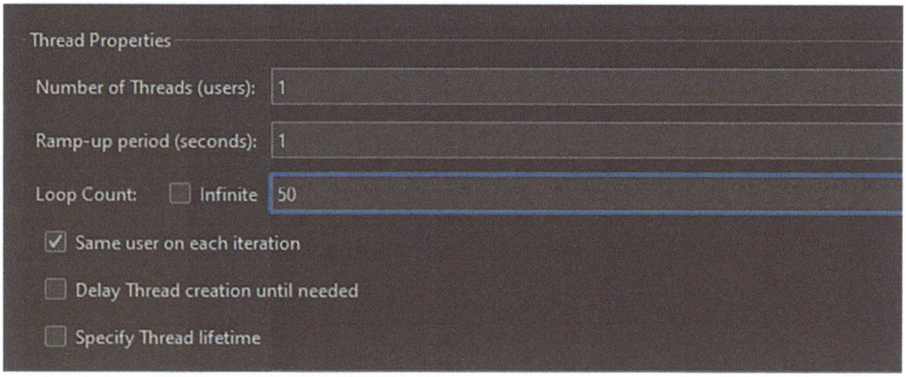

Fig. 5 Service request flow configuration

receive responses to transmitted requests and measure the delay between sending a request and receiving a response to a sent request. Requests will be sent to /POST 172.17.0.2/multipartfile.

The cloud application runs on an AWS virtual machine. JMeter accesses the application through the public DNS address of the load balancer, which in turn sends requests to the docker application in the EC2 instance. As part of the model, the application was launched in two regions. Applications that run in different regions are independent and have different addresses.

As part of the experiment, requests are made to three independent applications. Therefore, JMeter performs three independent "pools" of requests. The main goal is to measure network delays in the described three scenarios.

4 Results of the Network Delay Measurement Experiment in the Cloud and Edge Computing Model in High and Low Network Load Scenarios

The results of network delays (the time interval between sending a request and receiving a response) were obtained as a result of executing requests from one user in 50 iterations. In the process of generating the request, a POST request was sent with MultipartFile in the body of the HTTP protocol request. The server processes this MultipartFile and has returned an "It's ok" response if the request is successfully processed. Unlike cloud machines (2 CPU, 1GB RAM) the local machine has two times fewer computing resources (1 CPU, 512 MB RAM) for processing requests.

The latency level for all 50 generated requests is shown in Fig. 6.

It could be seen from Fig. 6 that the average level of network delays in the network model with edge computing is ~1015 ms. The peak latency is ~1023, and the minimum latency is 1008 ms.

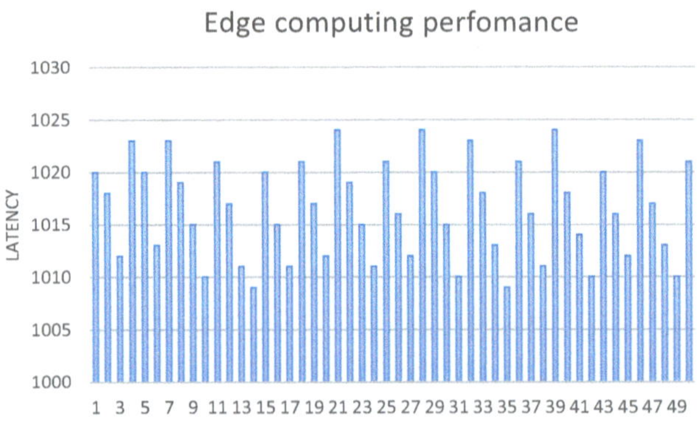

Fig. 6 Network latency with edge computing

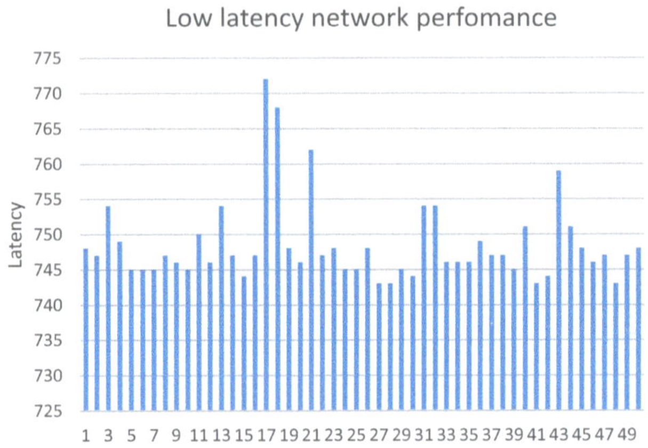

Fig. 7 Delay level in a low-load network

The following measurements consider the cloud computing model in a low-load network. This model simulates a network with few requests and a minimum load on the queues.

Figure 7 highlights the latency level for all 50 generated requests.

As can be seen from Fig. 7, the average network latency in the network model with cloud computing in a network with a low load is ~750 ms. The peak latency is ~772, and the minimum latency is 743 ms.

The next measurements consider a cloud computing model in a heavily loaded network. This model simulates a network with an excessive number of requests for the network resources (scenarios generating a large number of requests in one area).

The latency level for all 50 generated requests could be seen in Fig. 8. The average network latency in the network model with cloud computing in the high load network is ~1202 ms, the peak latency is ~1221, and the minimum latency is 1182 ms.

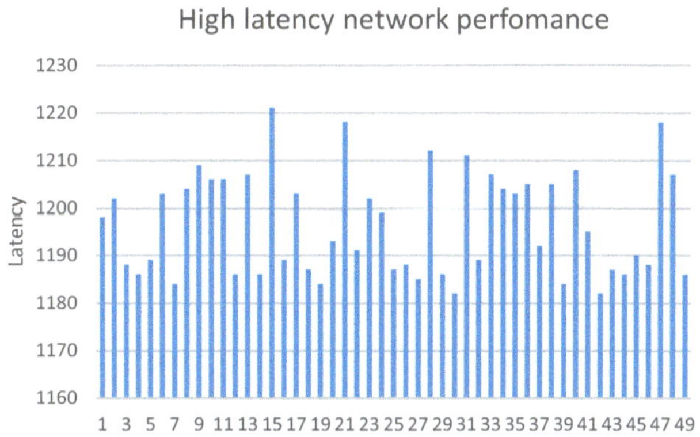

Fig. 8 Delay level in a high-load network

Table 1 Latency results in three types of networks

Network model/delay level	Minimum delay (ms)	Average delay (ms)	Max delay (ms)
Edge computing	1008	1015	1023
Cloud computing in a high-load network	1182	1202	1221
Cloud computing in a low-load network	743	750	772

The results, shown in Table 1, were received during the experiment. These results show the latency measurement in a network with edge computing, cloud computing in a high-loaded network, and cloud computing in a low-loaded network.

The experiment results in Table 1 highlight that edge computing only shows an advantage over cloud computing in a highly loaded network. Cloud computing in a low-load network shows better latency performance in comparison with the edge computing model.

Based on the received results, it can be argued that edge computing is more optimal to implement in networks with a high load. Smart city network high loads caused by high pedestrian and car traffic may require implementing the edge computing model.

5 Conclusion

5G technologies have a huge potential in the possible variety of services. The smart city is a concept of the future that will improve the quality of telecommunications services and the quality of life in general. It is a revision and a search for more effective analogs to previously used technologies required for the sustainable development of these services. Edge computing could become one of the analogs and significantly reduce network latency in a highly loaded network. Reducing latency will allow implementing the delay-sensitive services, which will beneficially affect the development of smart city applications.

Acknowledgments The research was not supported by any funding.

References

1. Muthanna, M.S.A., Wang, P., Wei, M., Rafiq, A., Josbert, N.N.: Clustering optimization of LoRa networks for perturbed ultra-dense IoT networks. Information. **12**, 76 (2021)
2. Muthanna, M.S.A., Muthanna, A., Rafiq, A., Hammoudeh, M., Alkanhel, R., Lynch, S., Abd El-Latif, A.A.: Deep reinforcement learning based transmission policy enforcement and multi-hop routing in QoS aware LoRa IoT networks. Comput. Commun. **183**, 33–50 (2021)
3. Muthanna, M.S.A., Wang, P., Wei, M., Abuarqoub, A., Alzu'bi, A., Gull, H.: Cognitive control models of multiple access IoT networks using LoRa technology. Cogn. Syst. Res., 62–73 (2021) ISSN 1389-0417
4. Chaaf, A., Saleh Ali Muthanna, M., Muthanna, A., Alhelaly, S., Elgendy, I.A., Iliyasu, A.M., et al.: Energy-efficient relay-based void hole prevention and repair in clustered multi-AUV underwater wireless sensor network. Secur. Commun. Netw. **2021** (2021)
5. Rafiq, A., Ping, W., Min, W., Muthanna, M.S.A.: Fog assisted 6TiSCH tri-layer network architecture for adaptive scheduling and energy-efficient offloading using rank-based Q-learning in smart industries. IEEE Sensors J. **21**(22), 25489–25507 (2021)
6. Muthanna, M.S.A., Alkanhel, R., Muthanna, A., Rafiq, A., Abdullah, W.A.M.: Towards SDN-enabled, intelligent intrusion detection system for Internet of Things (IoT). IEEE Access. **10**, 22756–22768 (2022). https://doi.org/10.1109/ACCESS.2022.3153716
7. Ahsan, R., Ping, W., Min, W., Ali, M.M.S., Nkerabahizi, J.N.: Mitigation impact of energy and time delay for computation offloading in an industrial IoT environment using Levenshtein distance algorithm. Secur. Commun. Netw., 1939–0114 (2022)
8. Aboulola, O., Khayyat, M., Al-Harbi, B., Muthanna, M.S.A., Muthanna, A., Fasihuddin, H., Alsulami, M.H.: Multimodal feature-assisted continuous driver behavior analysis and solving for edge-enabled internet of connected vehicles using deep learning. Appl. Sci. **11**, 10462 (2021)
9. Sharma, A., Cholda, P., Kumar, R., Dhiman, G.: Risk-aware optimized quickest path computing technique for mission-critical routing. Comput. Electr. Eng. **95**, 107436 (2021). https://doi.org/10.1016/j.compeleceng.2021.107436. (Scopus), (SCI) IF 3.818
10. Sharma, A., Kumar, R.: Computation of the reliable and quickest data path for healthcare services by using service-level agreements and energy constraints. Arab. J. Sci. Eng. **44**, 1–18 (2019). https://doi.org/10.1007/s13369-019-03836-4. (SCI) IF 2.334
11. Ashutosh Sharma, E., Podoplelova, G., Shapovalov, A.T., Tselykh, A.: Sustainable smart cities: convergence of artificial intelligence and Blockchain. Sustainability. (2021). https://doi.org/10.3390/su132313076. (SCI) IF 3.251

Organization of Smart City Services Based on Microservice Architecture

Mohammed Saleh Ali Muthanna, Dmitry Elkin, Semyon Likhtin, and Ammar Muthanna

1 Introduction

Recently, technologies have developed and become an integral part of our lives, and it is difficult to imagine modern life without the latest means or computer systems. We use all of the above in various spheres of life, such as medicine, industry, transport, and nuclear energy. These areas define critical processes that must be protected and constantly monitor the processed information or traffic transmitted inside the critical infrastructure [1].

Individual systems in critical infrastructure can be automated, for example, collecting Internet of Things data, monitoring a patient's condition remotely, or managing processes at a nuclear power plant. Systems that implement such processes can be called cyber-physical since they are an intermediate link between man and the world of machines [2].

In some cases, disruption of critical processes because of a failure or a centralized attack on cyber-physical systems in critical infrastructure can harm the health of others or damage an entire city or even a country. To avoid this, it is necessary to use a reliable info-communication system with suitable protection for a particular area of critical infrastructure. An important and essential factor in protecting such infrastructure is the need for constant monitoring of information or traffic within the network or system.

M. S. A. Muthanna (✉) · D. Elkin · S. Likhtin
Institute of Computer Technologies and Information Security, Southern Federal University, Taganrog, Russian Federation
e-mail: muthanna@sfedu.ru; delkin@sfedu.ru

A. Muthanna
The Bonch-Bruevich Saint-Petersburg State University of Telecommunications, St. Petersburg, Russian Federation

A. A. Abd El-Latif et al. (eds.), *Secure Edge and Fog Computing Enabled AI for IoT and Smart Cities*, EAI/Springer Innovations in Communication and Computing,
https://doi.org/10.1007/978-3-031-51097-7_3

Traffic interception, or traffic collection, is listening to a network interface using special libraries. The ability to intercept information can be used for both good and bad purposes. For example, it can be used by hackers for malicious purposes and by network administrators for monitoring, detecting looped traffic, detecting viruses and malware, etc.

As a result of the development of information technology and the traffic generated by people, there is also traffic generated by devices – Internet of Things traffic. It was necessary to bring this technology to one concept and standardize it. 5G/IMT-2020 is the latest standard in network architecture. This technology is very closely related to the "Internet of Things" because of one of the scenarios of its work – massive machine-type communications (mMTC).

The Internet of Things has several core applications. One of them is the "smart city." To manage urban systems, a smart city integrates info-communication technologies, including the Internet of Things [3]. This is an application of the Internet of Things, which aims to improve citizens' household communication, transport, and other conditions. However, it is worth noting that many services of this application can be implemented in rural areas. There are some problems in the "smart city" network organization, one of which is proposed to be solved in this work.

This chapter focuses on the effectiveness of traffic monitoring in an infrastructure that uses protocols that are part of the popular Transmission Control Protocol (TCP)/IP protocol stack. Such protocols allow transmitting information over wired and wireless communication channels. In addition to protocols for data transmission, there are various techniques for intercepting transmitted data.

2 Formulation of the Problem

This chapter aims to study the existing methods of organizing smart city services and develop an architectural solution with its subsequent implementation and testing on the IMT-2020 model network.

To achieve this goal, it is necessary to solve the following tasks:

- Study of existing software design methods.
- Selection of an effective architectural solution for the implementation of smart city services;
- Development of solution architecture.
- Development of a microservices interaction model to test the architecture of the solution.
- Development of service components.
- Implementation of the developed model on a laboratory stand.
- Conducting experiments to test the performance of the architectural solution.
- Analysis of the obtained results.

According to the task set, it became necessary to study and choose an effective architectural solution from all possible ones. Select software to implement this architecture, as well as deploy the solution itself and develop the necessary components for its operation. To identify the possibility of obtaining load results, to decide which indicators are important and which are not. Choose programming languages and libraries for developing modular solutions.

3 Methodology

This section comprises the architecture of the developed service organization software, service IoTDM, smart device traffic generator, statistics service, and the general architecture of the developed software.

3.1 The Architecture of the Developed Service Organization Software

To organize the monitoring architecture of the smart city system, the necessary task is to determine the type of architecture for deploying and developing the system. Based on the knowledge gained about the main architectures, it was concluded that the microservice architecture is most suitable for deploying the smart city system and monitoring it. The main reason for choosing this method of development was the need for constant scaling on the horizontal and vertical axes. For this reason, the monolithic method of organization instantly disappeared. The multitier architecture did not fit because of the centralized database. Serverless architecture is not suitable for economic reasons, most functions for working with a broker (publish/sent/receive) work constantly in an infinite loop, and with serverless architecture, first of all, the number of launches is paid, if the function works for more than 5 min, then it is not economically feasible [4].

As described earlier, the Message Queuing Telemetry Transport (MQTT) broker Eclipse Mosquitto will become the central device in the service. To expand the service with a broker, server no.1 was taken. The database, back end, service worker, and front end [5] will be located on the same server to monitor the load of the broker. Such an architectural decision in the location of services was made in connection with the need to obtain a minimum round-trip delay between the broker and the statistics service. Also, as the observation that the statistics service practically does not load the server RAM, but only takes system data from the broker, to save server resources, they were located in one place.

Figure 1 shows that each application is running in its own docker container, which allows you to manipulate it without affecting the others. Also, although the statistics service and the broker are running on the same server, they are running on

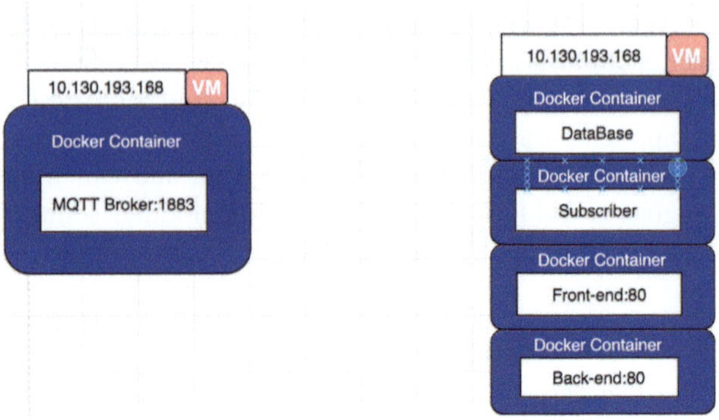

Fig. 1 MQTT broker and statistics service

entirely different virtual machines; this is done so that the load monitoring does not include load data from the statistics service, which contributes to obtaining cleaner statistics.

3.2 *Service IoTDM*

A necessary service for the management and control of Internet of Things (IoT) devices. First of all, IoTDM was used to create a model of "smart" devices, as well as to control these devices and then generate data on smart devices using the Internet of Things Data Management (IoTDM) and Application Programming Interface (API) [6].

The one Machine to machine (M2M) architecture, according to its specification, defines the physical things of the Internet of Things as "entities," just highlighting several of their types:

Applicatiton entity (AE) is the essence of the application; that is, not only the Internet of Things itself but also smartphones, personal computers, and other devices that have monitoring functions and the ability to control other devices fall under this definition.

1. Common services entity (CSE): A typical service object, in other words, a gateway for accessing IoT control and monitoring or the head control device itself.
2. Infrastructure node common services entity (in-CSE): The infrastructure of the common service object node is the senior object in the hierarchy. It is hosted in the cloud of the service provider.
3. Middle node (MN-CSE): A middle node located in the home domain, mapping a physical object to a resource tree, can be a gateway.
4. Container: Container provides a combination of data, which can be both parent and child (sub-container).
5. ContentInstance: Data generated by IoT devices, sensors, and so on.

Guided by the one M2M documentation, it is possible to develop own resource tree, which can be conveniently applied to micro- and macro-level architectures.

Fig. 2 Service IoTDM

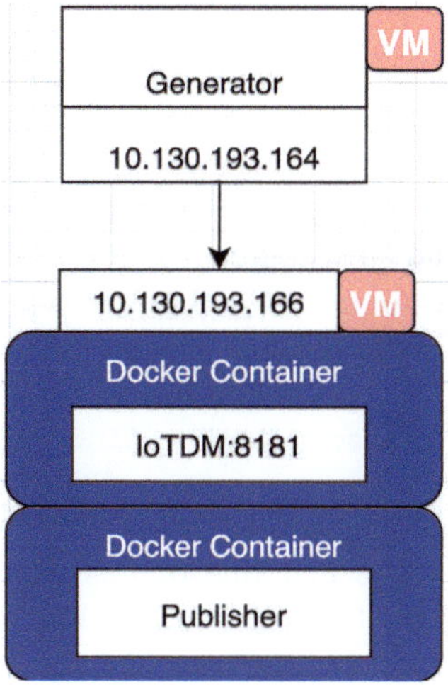

The IoTDM service can also be a broker of "smart" devices and provide monitoring and the ability to configure access policies to the Internet of Things resources [7]. However, when using a microservice architecture, it became necessary to separate "smart devices" and monitoring and analytics systems.

When using an MQTT broker, "smart" devices can be connected directly to the broker. However, with a large number of devices of different types, deployment problems arise since it is necessary to connect each device manually and configure the publication of data from the device to the broker's "topic."

In the case of using the IoTDM service, you can first group "smart" devices. Devices can be grouped by their type – for example, group temperature, touch, and other sensors together. With this structure, it becomes possible to publish data immediately from a group of devices. IoTDM can store data from devices, which is essential when there is a temporary lack of communication between the device and the MQTT broker. After restoring the connection, the data will be published to the desired channel.

As shown in Fig. 2, the IoTDM service was deployed inside a docker container on server #2. Port 8181 was also "forwarded," which is used for API requests to the IoTDM service, necessary for manipulation, data collection, and other functions.

In addition to the IoTDM container, this server also hosts a small application for publishing data from IoTDM to the broker's MQTT topic. This application, using the IoTDM API, fetches data from the IoTDM "storage" and publishes it in a topic, an example of fetching data using the Python 2.7 programming language from a server running in an IoTDM container:

```
url1 = 'http://10.130.193.166:8282/Central_District_SPB/switch2/
municipality_78/CrossRoads1/Light1/Humm_Base/Hummidity_1'
c = iotdm_api.retrieve(url1)
```

In the response, we receive a JSON object containing the data of the "smart device."

To publish data to MQTT, the broker also needs a library running in the Python 2.7 programming language so that data collection and publication take place within the same microservice without adding an additional database to avoid increasing round-trip latency. Paho MQTT Client in Python, also developed by the Eclipse studio, was chosen as this library, which means 100 percent support with a broker and open source.

An example of publishing data to a broker after receiving a JSON object from a service IoTDM:

```
client1 = paho.Client("iotdm1")
sensor_data = find_key(c[1], "con")
                    publish_key = "Ligth1/hum1"
                    ret = client1.publish(publish_key, sensor_data)
```

The first line of this example uses the paho mqtt library call in the client1 variable. In the second line of this example, we find the "con" key in the JSON object, it contains the data from the IoT device. In the next line, enter the publish_key constant and assign it the value of the topic in which we will publish information from this sensor. In the last line, the publish method of the client1 object is called passing the topic and data from the IoT device into it as parameters.

3.3 Smart Device Traffic Generator

In addition to the IoTDM service, a traffic generator is also noted in Fig. 2. For the model to work correctly, it is necessary to generate traffic by the created IoT devices, thereby creating a load on the MQTT broker. To implement the traffic generator, the developments made in the laboratory of software-defined networks were borrowed.

3.4 Statistics Service

In addition to the usually necessary applications for monitoring, the back end of the service, the front end of the service, and the database for monitoring, it was necessary to deploy an auxiliary service on the statistics server to collect system data about the broker. The so-called system subscriber is a small application subscribed to the base topic "SYS", an MQTT broker. To implement this application, the paho mqtt library, previously described [8], was taken.

The task of this application is to subscribe to the broker's system feed, collect load data, and store it in the database. An example of subscribing to a broker and publishing data in a database:

```
client = mqtt.Client()
client.on_connect = on_connect
client.on_message = on_message    client.connect("10.130.193.
168", 1883, 60)
def on_message(client, userdata, msg):
bytes = float(str(msg.payload)[2:-1])
```

```
cur.execute(query(connections_one_min), (bytes, datetime.
datetime.now()))
```

The first line is already familiar, as described above; here the client is defined as a method of the paho.mqtt class; below are the definitions of the on_connect method as a function of the on_connect application, similar to on_message. Next, the definition of the on_message function, which parses the string that comes in bytes and a message and saves it in the database.

3.5 The General Architecture of the Developed Software

When developing a model for load testing, it is necessary to consider the possibility of reducing the load by implementing additional services.

As shown in Fig. 3, an architectural solution that will reduce the load on the MQTT broker, it was proposed to implement a second MQTT broker with a

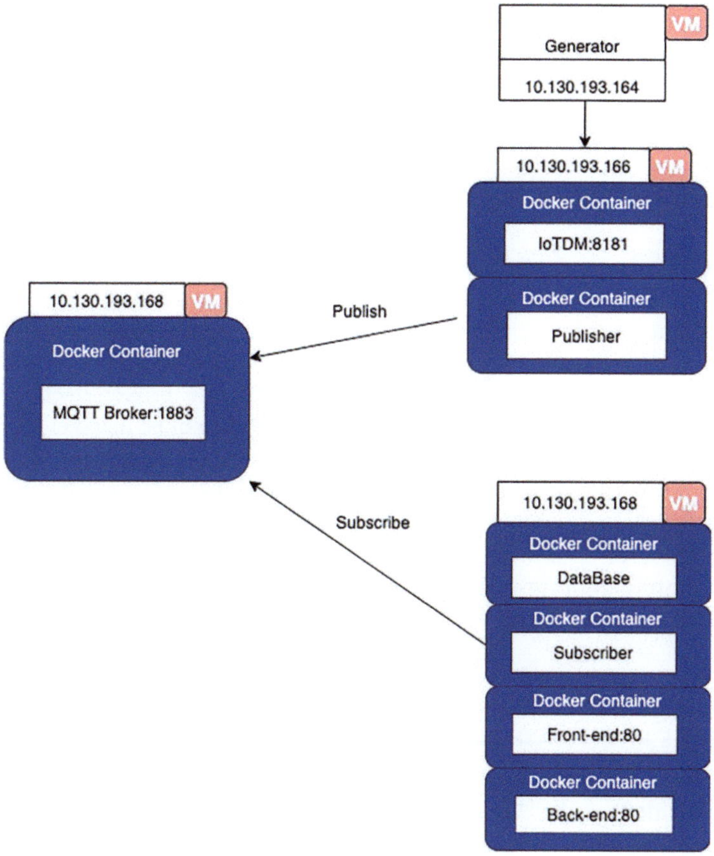

Fig. 3 General architecture of microservices

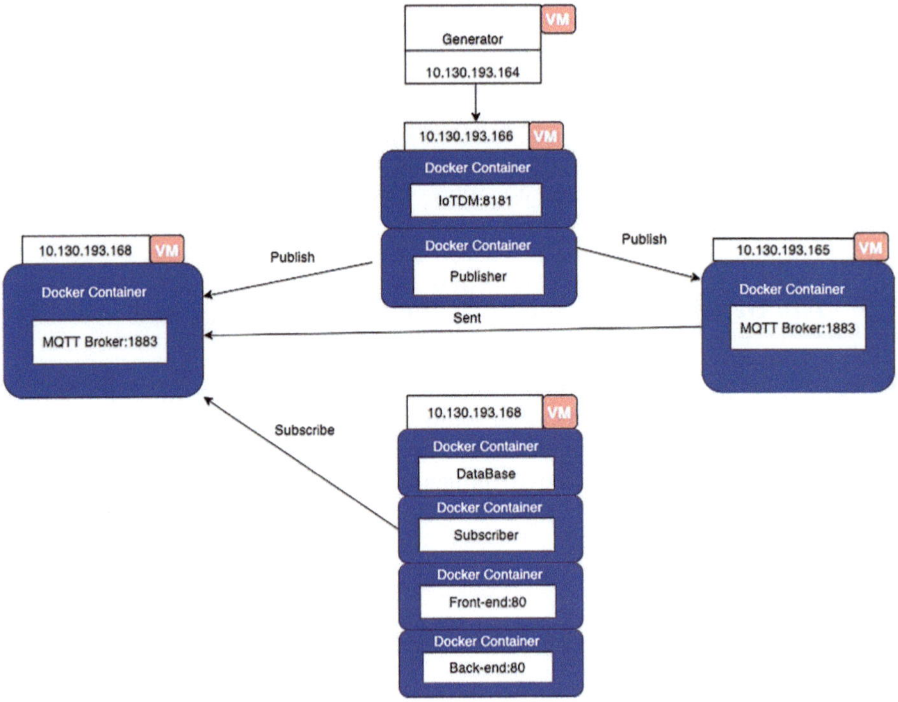

Fig. 4 General architecture of microservices with two brokers

bridged connection to the first one. This solution reduces the number of publications made by "smart devices" to the broker's topic, and if it is necessary to obtain data that came to the second broker, it is necessary and sufficient to organize a bridge connection between brokers and specify the necessary topics for publication in the first broker in the configuration settings of the broker itself. With this configuration, the information received in the topic will not be listed as "published" but will be listed as "received," and here you can also configure how often the data will be transmitted, which can significantly reduce the load on the main broker. Among other things, we can restrict user access to the information received at the machine level – allowing you to subscribe only to secondary brokers and publishing information in them that is necessary for a specific user group, which will significantly increase the security of the service as shown in Fig. 4.

4 Results and Discussion

This chapter defines the description of the testing process. When developing the model, the possibility of using the Eclipse Mosquitto broker as a central system for the smart city service network was investigated, and the possibility of distributing sensors across servers, in this case, IoTDM servers, was also investigated.

4.1 Results of Testing the Developed Model Without Load Distribution Between Microservices

During testing, no results were obtained on the CPU load, as well as the percentage of RAM utilization, due to the incredibly low load. The reason for this was a quantitative lack of traffic on the generator, a small number of publications in the broker's topic, which made it impossible to compare the load in different experiments.

For the above reason, it was decided to compare quantitative data on publications made, as well as received and sent packets in bytes. To obtain these test results, a web interface was designed.

In total, it was necessary to display three graphs, two of which are located together, for a visual demonstration of the difference. Figure 5 shows the quan-

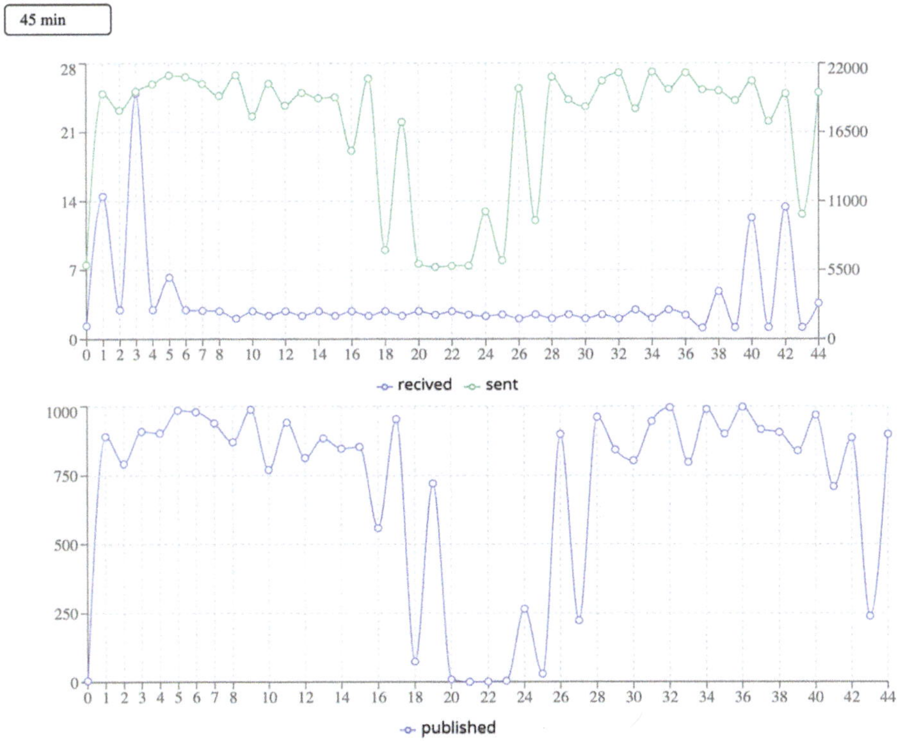

Fig. 5 Designed interface for a visual demonstration of the load

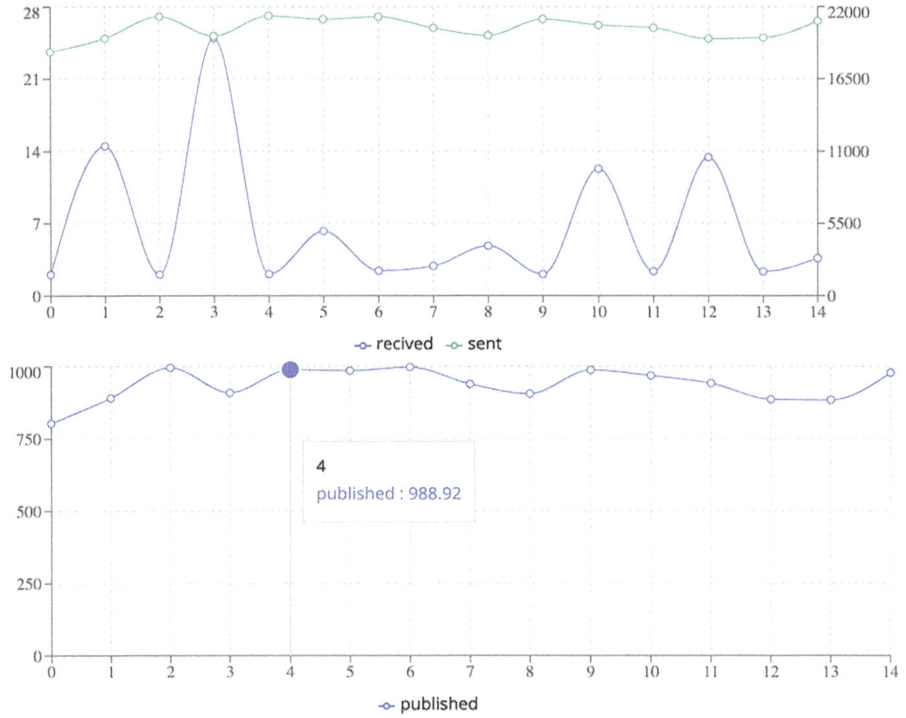

Fig. 6 Maximum value of publications in bytes without load balancing

titative values of sent (green graph) and received (purple graph) in bytes. It is immediately worth noting that the value of the received bytes is negligible due to the lack of transmission of anything to the broker (the value of publications does not count). The number of bytes sent is larger than the received ones due to the constant removal of statistics in the endless loop mode. The lower axis is the number of minutes elapsed since the beginning of the experiment.

Figure 6 shows a graph of the dependence of the number of packets sent or received from publications – this is the maximum possible number of bytes in a given hardware configuration (a model without load distribution between microservices). As can be seen from the graph, the maximum number of bytes was 1000 bytes/min.

4.2 Results of Testing the Developed Model with Load Distribution Between Microservices

The second stage of testing was the connection of the second MQTT broker in bridge mode to the first one. The main task was to reduce the published information in the first broker by redirecting the publication of some sensors to the second one.

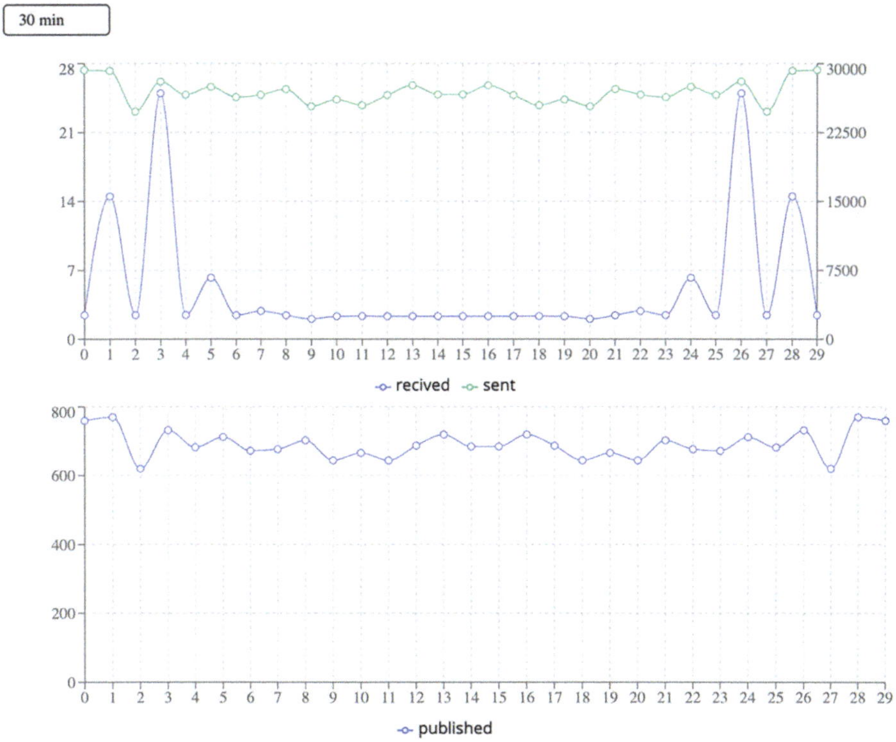

Fig. 7 Maximum load when implementing the second broker in bridge mode

Figure 7 shows the same graph, but this time we look at a model with load distribution across microservices.

The maximum amount of published information on the first broker, when transferring data publications from sensors to the second one, was ~800 bytes/min. It is also worth noting that the load was not distributed equally 50/50; this is only due to the fact that the task was to show the possibility of reducing the number of publications. Also, the second broker did not transmit information to the first of the sensor data topics, which did not affect the number of bytes received on the first broker.

5 Conclusion

In this work, a microservice architecture was selected and implemented. A model of the smart city application was constructed using this architecture. MQTT broker, Eclipse Mosquitto, was chosen as the central service for the microservice architecture, and IoTDM was chosen as the service for grouping sensors. A broker load monitoring system was designed and implemented, and a traffic generation service and a service for publishing sensor data. The architecture was validated, demonstrating the possibility of reducing the broker's load. Based on the testing results, appropriate recommendations were developed, identifying the architecture's strengths and weaknesses. Additionally, a Web interface has been implemented to allow for graphical monitoring of the Eclipse Mosquitto broker's load. The source code for the Eclipse Mosquitto broker has been made publicly available.

Acknowledgement The research is supported by postdoc fellowship granted by the Institute of Computer Technologies and Information Security, Southern Federal University, project № P.D./22-01-KT.

References

1. Muthanna, M.S.A., Wang, P., Wei, M., Rafiq, A., Josbert, N.N.: Clustering optimization of LoRa networks for perturbed ultra-dense IoT networks. Information. **12**, 76 (2021)
2. Muthanna, M.S.A., Muthanna, A., Rafiq, A., Hammoudeh, M., Alkanhel, R., Lynch, S., Abd El-Latif, A.A.: Deep reinforcement learning based transmission policy enforcement and multi-hop routing in QoS aware LoRa IoT networks. Comput. Commun. **183**, 33–50 (2021)
3. Muthanna, M.S.A., Wang, P., Wei, M., Abuarqoub, A., Alzu'bi, A., Gull, H.: Cognitive control models of multiple access IoT networks using LoRa technology. Cogn. Syst. Res., 62–73 (2021) ISSN 1389-0417
4. Chaaf, A., Saleh Ali Muthanna, M., Muthanna, A., Alhelaly, S., Elgendy, I.A., Iliyasu, A.M., et al.: Energy-efficient relay-based void hole prevention and repair in clustered multi-AUV underwater wireless sensor network. Secur. Commun. Netw. **2021** (2021)
5. Rafiq, A., Ping, W., Min, W., Muthanna, M.S.A.: Fog assisted 6TiSCH tri-layer network architecture for adaptive scheduling and energy-efficient offloading using rank-based Q-learning in smart industries. IEEE Sensors J. **21**(22), 25489–25507 (2021)
6. Muthanna, M.S.A., Alkanhel, R., Muthanna, A., Rafiq, A., Abdullah, W.A.M.: Towards SDN-enabled, intelligent intrusion detection system for internet of things (IoT). IEEE Access. **10**, 22756–22768 (2022). https://doi.org/10.1109/ACCESS.2022.3153716
7. Ahsan, R., Ping, W., Min, W., Ali, M.M.S., Nkerabahizi, J.N.: Mitigation impact of energy and time delay for computation offloading in an industrial IoT environment using Levenshtein distance algorithm. Secur. Commun. Netw., 1939–0114 (2022)
8. Aboulola, O., Khayyat, M., Al-Harbi, B., Muthanna, M.S.A., Muthanna, A., Fasihuddin, H., Alsulami, M.H.: Multimodal feature-assisted continuous driver behavior analysis and solving for edge-enabled internet of connected vehicles using deep learning. Appl. Sci. **11**, 10462 (2021)

Pseudo-Random Error-Correcting Codes in Network Coding

Sergey Vladimirov

1 Introduction

The development of modern communication networks of the fifth and sixth generations and the emergence of new communication services lead to increased requirements for content delivery and data transmission systems and pose an important task for researchers and developers—reducing the time of packet transmission between network nodes [6, 7, 30]. But with the growth in the number and quality of services, both the number of subscribers and the number of service sources grow, which ultimately leads to a significant increase in traffic transmitted over the network and, accordingly, an increase in transmission delays caused by the need to process and route traffic [14, 22]. When considering this set of tasks in the complex, there is an obvious contradiction, which is intended to be eliminated by the network coding mechanism.

Network coding is a formation of linear combinations of data packets at intermediate nodes of a data transmission network and the subsequent recovery of these packets at destination nodes [1, 2, 16, 31]. As a rule, network encoding of data packets is reduced to a reversible linear transformation of two or more data packets through mathematical operations. In systems with network coding, a data transmission network is usually represented as a finite directed graph, in which the nodes correspond to individual network devices, and the edges correspond to the data transmission channels connecting them. Graph nodes are divided into source nodes—only with output edges; destination nodes—only with input edges; internal or transit nodes that have input and output edges and provide transit and related processing of data transmitted from the source to the recipient. Network encoding

S. Vladimirov (✉)
St. Petersburg State University of Telecommunications, St. Petersburg, Russian Federation

A. A. Abd El-Latif et al. (eds.), *Secure Edge and Fog Computing Enabled AI for IoT and Smart Cities*, EAI/Springer Innovations in Communication and Computing,
https://doi.org/10.1007/978-3-031-51097-7_4

can reduce the receive latency and total number of transmitted packets for multicast transmission in wired and wireless networks and increase network performance [4, 16, 20, 23, 26].

2 Principle of Network Coding

The main mathematical operation used in network coding systems is bitwise addition modulo 2, as the simplest reversible operation. It is also possible to use other reversible operations that provide unambiguous decoding, for example, mathematical operations on finite Galois fields [16, 23, 27].

Consider as an example the case of two sources. Take two packages P_1 and P_2. In the simplest case, the result of their network encoding is represented by a packet

$$P_{NC} = P_1 \oplus P_2, \tag{1}$$

where $\oplus$ is the bitwise addition modulo 2 operator, also known as the "exclusive-OR" (XOR) operation.

Thus, knowing the encoded packet P_{NC} and one of the original packets, it is possible to recover the second original packet:

$$\begin{cases} P_2 = P_1 \oplus P_{NC} = P_1 \oplus P_1 \oplus P_2 = P_2; \\ P_1 = P_2 \oplus P_{NC} = P_2 \oplus P_1 \oplus P_2 = P_1. \end{cases} \tag{2}$$

When considering the mechanism of network coding, one traditionally talks about transmission in a "butterfly" topology network, shown in Fig. 1 [2, 16]. The network topology contains two source nodes S_1 and S_2, which must simultaneously transmit data packets P_1 and P_2 to destination nodes R_1 and R_2 at the same time. Routing of data packets from source nodes to destination nodes is provided by six transit routing nodes from G_A to G_F.

Consider the data transfer procedure in the butterfly network according to the scenario of simultaneous data transfer. The numbers near the arrows in Fig. 1 conditionally show the order in which packets pass through the network.

In normal multicast transmission without network encoding, P_1 and P_2 packets do not arrive at nodes R_1 and R_2 at the same time. Node R_2 receives both packets conditionally after five steps, and node R_1—after six, i.e. a delay occurs. The reason is the fact that in the route G_C–G_D packets are transmitted sequentially. In the case of network coding the G_C node calculates a linear combination P_{NC} by (1) and immediately transmits it to G_D. Therefore, nodes R_1 and R_2 receive both packets simultaneously in five steps.

Thus, the use of network coding allows one to win the transmission time of one packet and reduce the amount of traffic transmitted in the G_C–G_D route approximately by half. When accurately calculating the volume reduction of transmitted traffic, it is necessary to take into account the overhead and sizes of the protocol headers which are necessary to ensure the operation of the network encoding and packet routing mechanisms.

3 Error Control in Network Coding

A feature of the system with network coding is the increased requirements for the noise immunity of transmission channels connecting the nodes of the data transmission network [12].

Consider the butterfly network (Fig. 1). Assume that on the route G_A–G_E there was an error E_{AE} injected into the transmitted packet P_1 so that the message $P_1' = P_1 \oplus E_{AE}$ is received at the input of node G_E. Thus, recipient R_1 will not be able to receive packet P_1 or packet P_2 because

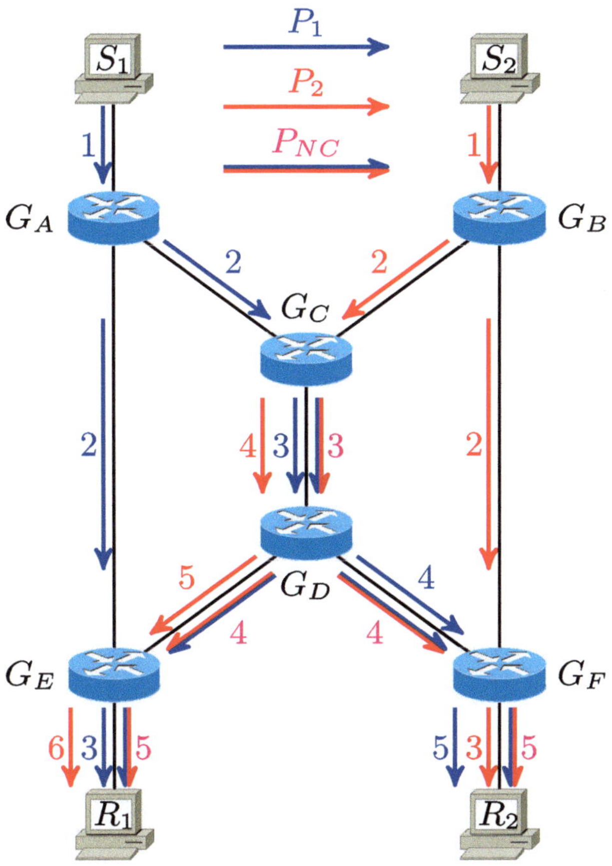

Fig. 1 Multicast and network coding in butterfly network

$$P_{NC} \oplus P_1' = P_1 \oplus P_2 \oplus P_1 \oplus E_{AE} = P_2 \oplus E_{AE}. \tag{3}$$

In the event of an error injection on any of the G_A–G_C, G_B–G_C, or G_C–G_D routes, the destination nodes R_1 and R_2 will not be able to receive packets P_1 and P_2, respectively.

Assume that while sending message P_1 along the route G_A–G_C an error E_{AC} was injected into it. Accordingly, the node G_C received the message $P_1'' = P_1 \oplus E_{AC}$. As a result of the linear combination of messages, node G_C then transmitted the packet $P_{NC}'' = P_1'' \oplus P_2$, which was received by the receiving nodes R_1 and R_2. Thus, when restoring messages, node R_1 will receive a message

$$P_{NC}'' \oplus P_1 = P_1'' \oplus P_2 \oplus P_1 = P_2 \oplus E_{AC}, \tag{4}$$

and node R_2 will receive the packet

$$P_{NC}'' \oplus P_2 = P_1'' \oplus P_2 \oplus P_2 = P_1 \oplus E_{AC}. \tag{5}$$

Therefore, an error injected into the transmitted messages on any of the routes will result in total or partial data loss [12].

It should be noted that network coding can lead to error propagation. If in the examples, considered above, both source packages were injected by errors:

$$\begin{cases} P_1' = P_1 \oplus E_1; \\ P_2' = P_2 \oplus E_2, \end{cases} \tag{6}$$

then the resulting package P_{NC}' will contain errors from both source packages:

$$P_{NC}' = P_1' \oplus P_2' = P_1 \oplus E_1 \oplus P_2 \oplus E_2 = P_{NC} \oplus E_1 \oplus E_2. \tag{7}$$

Thus, when implementing systems with network coding, it is required to solve the problem of effective error control [12]. To do this, it is necessary either to use communication channels that ensure data transmission without errors, which is often impossible, or to use systems with error detection and/or forward error correction. Simple error detection allows error messages to be requeried through the feedback channel if necessary. Forward error correction allows to correct some of the errors up to a certain multiplicity (depending on the code used) and thereby avoid re-requests.

That is why, one of the most important directions in the study of systems with network coding is the question of their use in conjunction with error-correcting codes. Such research is carried out around the world by various groups of scientists and developers.

Since 2009, researchers from MIPT have been studying the use of Gabidulin rank codes for systems with random network coding [11, 17, 18].

In 2011, a group of developers from the American R&D company Inter-Digital [24] presented a work on the use of binary and ternary convolutional codes in radio broadcast systems with physical network coding.

In 2013, a team of researchers from MIT [5] considered the application of network coding in conjunction with channel convolutional coding for low power wireless sensor networks.

Many researchers [8, 10, 25] suggest using non-binary error-correcting Reed–Solomon codes for transmission systems with network coding. Such codes provide high corrective power but have greater computational complexity compared to simple binary codes.

Another example of efficient complex error-correcting codes popular among researchers is low-density LDPC codes [13, 15, 21, 28, 29].

Thus, the study of the use of error-correcting codes in network coding is an important and urgent scientific task. This paper considers the use of pseudo-random error-correcting codes based on simple and composite pseudo-random sequences in network coding.

4 Using Binary Pseudo-Random Codes to Combat Errors in Network Coding

Pseudo-random binary error-correcting codes are sets of binary pseudo-random sequences (PRS). Among them, there are codes based on simple PRS, such as maximum length sequences or Barker codes, and codes based on composite PRS, such as Gold, Kasami and Gordon–Mills–Welch codes (sequences).

The property of composite PRSs is that they are formed from simple PRS (usually MLS), by adding them with bitwise addition modulo 2, which corresponds to the network coding procedure (1).

As an example of codes based on binary PRS for network coding, consider the maximum length code (MLC) and the Gold code.

4.1 Maximum Length Code

Equidistant cyclic (n, k) maximum length codes (MLC) are constructed on the basis of maximum length sequences (MLS)—pseudo-random sequences of maximum period over a finite Galois field GF(2^k) generated by irreducible primitive polynomial $p(x)$. The length n of the MLC codeword over the field GF(2^k) is equal to $n = 2^k - 1$ [9].

To form the MLS, either a hardware circuit or a software algorithm is used.

The hardware circuit is implemented on the basis of a linear feedback shift register, which is constructed based on the form of the generating polynomial of the Galois field $p(x)$. Consequently, the polynomial $p(x)$ can also be considered a generating polynomial of both the MLS and the corresponding MLC.

In the program algorithm, the MLS is considered as a recurrent sequence $\{s\}$ calculated by the trace function $Tr(x)$ of the Galois field element [3] by the formula

$$\{s\} = \left[Tr(c)Tr(c\epsilon)Tr(c\epsilon^2)\ldots Tr(c\epsilon^{2^k-2})\right], \tag{8}$$

where $c = \epsilon^z$ is the initial phase of the MLS, which is an element of the Galois field GF(2^k).

The trace function $Tr(x)$ for the element ϵ^i of the field GF(2^k) is calculated by the formula [3]

$$Tr(\epsilon^i) = \epsilon^i + (\epsilon^i)^2 + (\epsilon^i)^{2^2} + \cdots + (\epsilon^i)^{2^{k-1}}. \tag{9}$$

The set of MLSs $\{s\}$ constructed over the Galois field GF(2^k) for all possible initial phases forms a MLC (n, k). This code can be considered as systematic, assuming that the information elements are the first k elements of the sequence, and as non-systematic—in this case, the information elements are the initial phase c of the MLS. The minimum code distance of such a code is $d_{\min} = 2^{k-1}$.

As an example, consider a MLS of length $n = 15$ over the field GF(2^4). It forms a binary MLC $(15, 4)$ with $d_{\min} = 8$. This code is guaranteed to fix $t \leq 3$ errors.

4.2 Gold Code

The Gold code is formed on the basis of composite pseudo-random Gold sequences $\{s\}_G$, which in turn are calculated by modulo 2 addition of two MLSs $\{s\}_1$ and $\{s\}_2$ with the same n over different Galois fields GF(2^k) formed by the polynomials $p_1(x)$ and $p_2(x)$, respectively:

$$\{s\}_G = \{s\}_1 \oplus \{s\}_2. \tag{10}$$

The resulting Gold code has the dimension (n, k_G), where $k_G = 2k$. Thus, the number of information elements in the codeword of the Gold code is equal to twice the number of information bits of the MLCs from which it is formed. So, the Gold codeword carries information of both MLCs's codewords, from which it is composed [19].

The MLC example $(15, 4)$ considered above corresponds to the Gold code $(15, 8)$, which has $d_{\min} = 3$ and is guaranteed to correct only single errors ($t = 1$).

4.3 Decoding of MLC and Gold Code

The decoding of the MLC and the Gold code can be done in various ways. The method of majority decoding based on the dual basis of the Galois field is used in this paper. In this method, all information elements are calculated for each k-digit block of a codeword closed in a ring. The principle of closing the codeword into a ring and selecting all n k-digit blocks is shown in Fig. 2 using the example of (15, 4) MLC. Enumeration of all n k-digit blocks allows organizing majority decoding to correct errors that occur when data is sent over a communication channel. At the same time, the use of the majority principle makes it possible to correct bursts of errors with a multiplicity greater than t.

The general principle of error correction by this method is shown in Fig. 3. The algorithm considers the decoding of the codeword M of the (n, k)-code carrying the information word D with values from 0 to $2^k - 1 = n$.

The time for decoding codewords of MLC and the Gold code, and hence the delay in receiving codewords, depend on the number of errors contained in the received codeword and the method of implementing the decoding device. According to the algorithm shown in Fig. 3, in the general case, decoding a codeword requires enumeration of all n k-digit blocks. Taking into account that the algorithm is based on the principle of majority, to make a decision about the unambiguity of decoding, it is sufficient to obtain more than $n/2$ identical results (checking the $(a_D > n/2)$ condition in Fig. 3). Accordingly, codeword decoding can be completed after $\frac{(1+n)}{2}$ k-digit blocks has been iterated. This situation occurs either in the case of decoding an error-free codeword or in case when errors are placed in the last bits of the codeword. The number of initial error-free bits depends on the type and size of the code. For example, for MLC (15, 4), it may be sufficient to process the first eight 4-digit blocks. This case is possible when the errors are localized in the last 3 bits of the codeword as shown in Fig. 2. For the Gold code (15, 8), early decision is possible only in the case of an error-free codeword.

To estimate the probabilistic characteristics of MLC and the Gold code, modeling was carried out according to the scheme in Fig. 4.

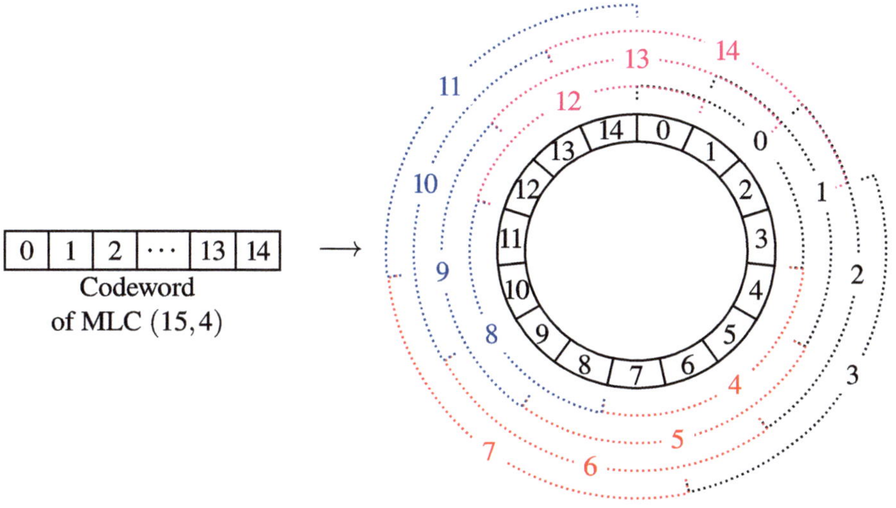

Fig. 2 Cyclizing of MLC (15, 4) codeword and highlighting of n k-digit blocks

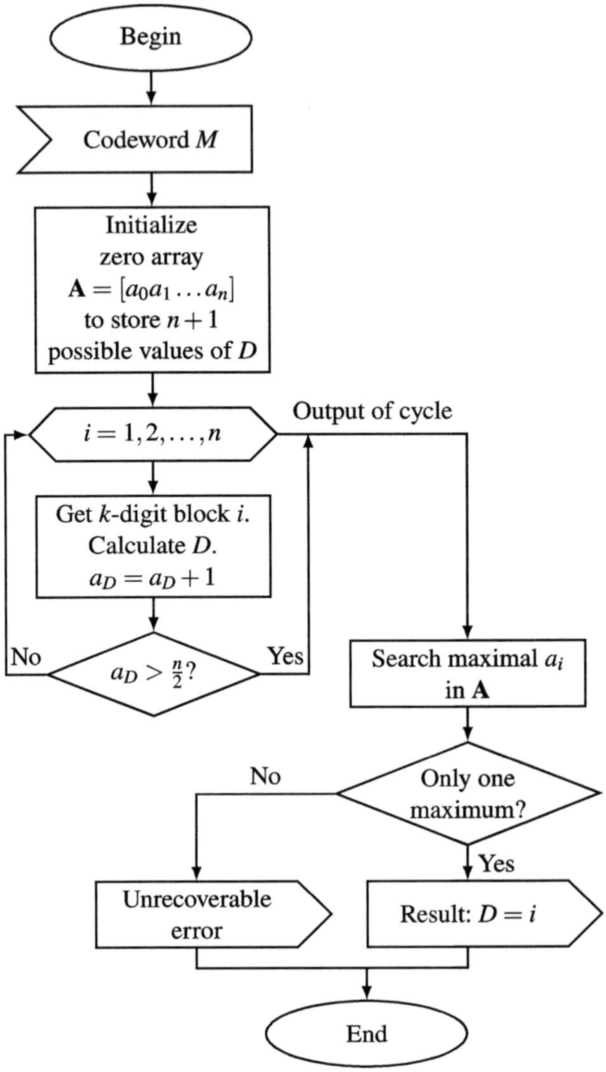

Fig. 3 Majority decoding of MLC or Gold code

Fig. 4 The scheme of model to estimate the probabilistic characteristics of MLC and the Gold code

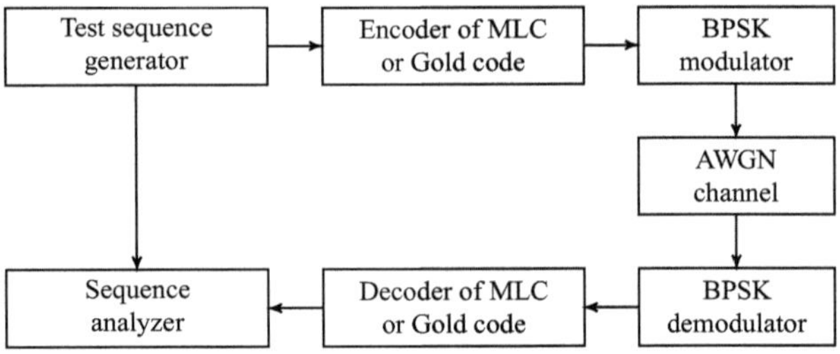

The plots in Fig. 5 show the probabilistic characteristics of the (15, 4) MLC and the (15, 8) Gold code during transmission over the AWGN channel for the case of BPSK modulation. The probabilities of correct decoding of the received codeword (P_{CD}), incorrect decoding (P_{ID}), and detected unrecoverable error (P_{UE}) are given.

4.4 Network Coding with MLC and Gold Code

Network coding with MLC and Gold code in butterfly network is shown in Fig. 6.

Sources S_1 and S_2 encode the transmitted data D_1 and D_2 with different MLCs of the same dimension, obtaining codewords M_1 and M_2, respectively. Router nodes G_A and G_B forward these codewords to encoder node G_C and destination nodes R_1 and R_2. At node G_C, from codewords M_1 and M_2, a Gold codeword $G = M_1 \oplus M_2$ is formed, which, after passing through node G_D, arrives at the destination nodes, which decode and restore initial data D_1 and D_2.

A feature of the proposed network coding method is that the codeword G of the Gold code contains both blocks of initial data D_1 and D_2. Therefore, if any of the codewords M_1 and M_2 are lost on the network routes G_A–R_1 and G_B–R_2, respectively, the data will be received and restored by receivers R_1 and R_2, without requiring a re-query of both data blocks. This leads to a reduction in data transmission delays at the expense of increased data coding redundancy.

Decoding of received codewords at receivers R_1 and R_2 can be done in three different ways. Let us consider them using the node R_1 as an example.

The first method involves the use of the standard for network coding method of summing the received codewords M_1 and G to obtain the codeword $M_2 = M_1 \oplus G$. When summed, the received codeword M_2 will contain errors from both received words. To reduce the number of errors before summation, the codeword M_1 should be decoded with error correction.

The second way is to decode codewords M_1 and G without restoring codeword M_2. The probabilistic characteristics of the MLC and the Gold code presented in Fig. 5 show that the probability of receiving correctly decoded source blocks D_1 and D_2 at the same time will not exceed the probability of correct decoding for the Gold code. This method may provide additional error checking. If at node R_1 different values for data block D_1 are obtained during decoding of MLC and Gold code codewords, then this indicates an unrecoverable error. Therefore, received codewords must be re-requested on the feedback channel or discarded if re-request fails or disabled.

The third method takes into account the fact that MLC has a greater corrective ability than the Gold code. Accordingly, with comparable quality of transmission channels (comparable signal-to-noise ratio) and different decoding results of M_1 and G, we can assume that the decoding result of M_1 will be more correct. Hence, based on the postprocessing of the codeword G according to the result of decoding M_1, it becomes possible to correct errors that occurred during transmission over the channel.

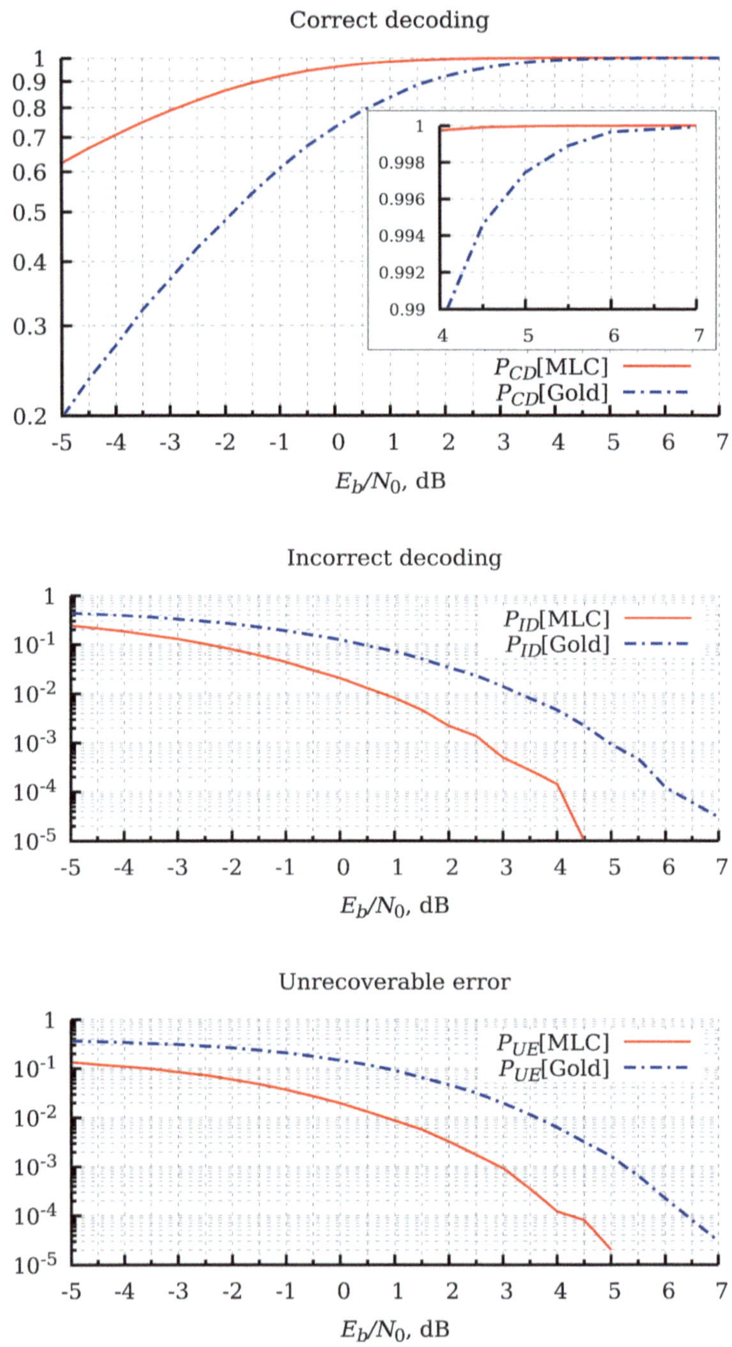

Fig. 5 Probability measures of MLC and Gold code

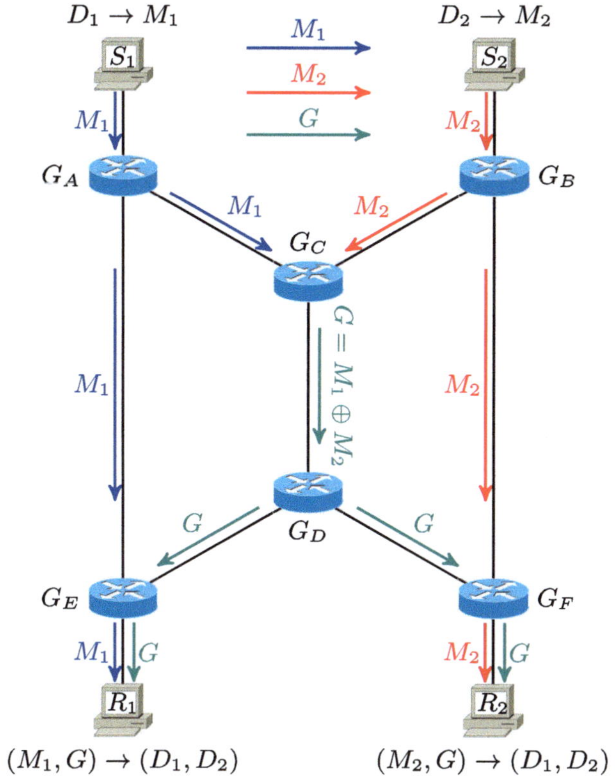

Fig. 6 Network coding with MLC and Gold code in butterfly network

Let us consider the order of postprocessing using the example of the destination node R_1 using MLC (15, 4) and the Gold code (15, 8). The block diagram of the decoding method with postprocessing is shown in Fig. 7.

Upon completion of the reception, node R_1 decodes codewords M_1 and G, receiving data blocks D_{1M}, D_{1G} and D_{2G}. If D_{1M}, and D_{1G} match, then we assume that the data blocks were received correctly. Otherwise, we assume that $D_1 = D_{1M}$ is true, and G contains errors. We correct errors in G until the results of D_{1G} and D_{1M} match. To do this, we sequentially consider all 8-digit blocks of the codeword G closed in a ring. For each series, we calculate D_{1G}, comparing it with D_{1M}. If there is a mismatch, we go through all the digits of the 8-digit block and for each of them we check whether its inversion will affect the result of the comparison. If D_{1G} becomes equal to D_{1M} when a bit is inverted, then we consider that there was an error in this bit and save the correction. Similarly, we consider all the following 8-digit blocks until all the series of the codeword G closed in a ring give the result $D_{1G} = D_{1M} = D_1$ during decoding. After that, we calculate the errors in G corrected and calculate $D_2 = D_{2G}$. If, after double sequential enumeration of all 8-digit blocks, errors are not corrected, we consider such an error unrecoverable and output the corresponding decoding failure signal.

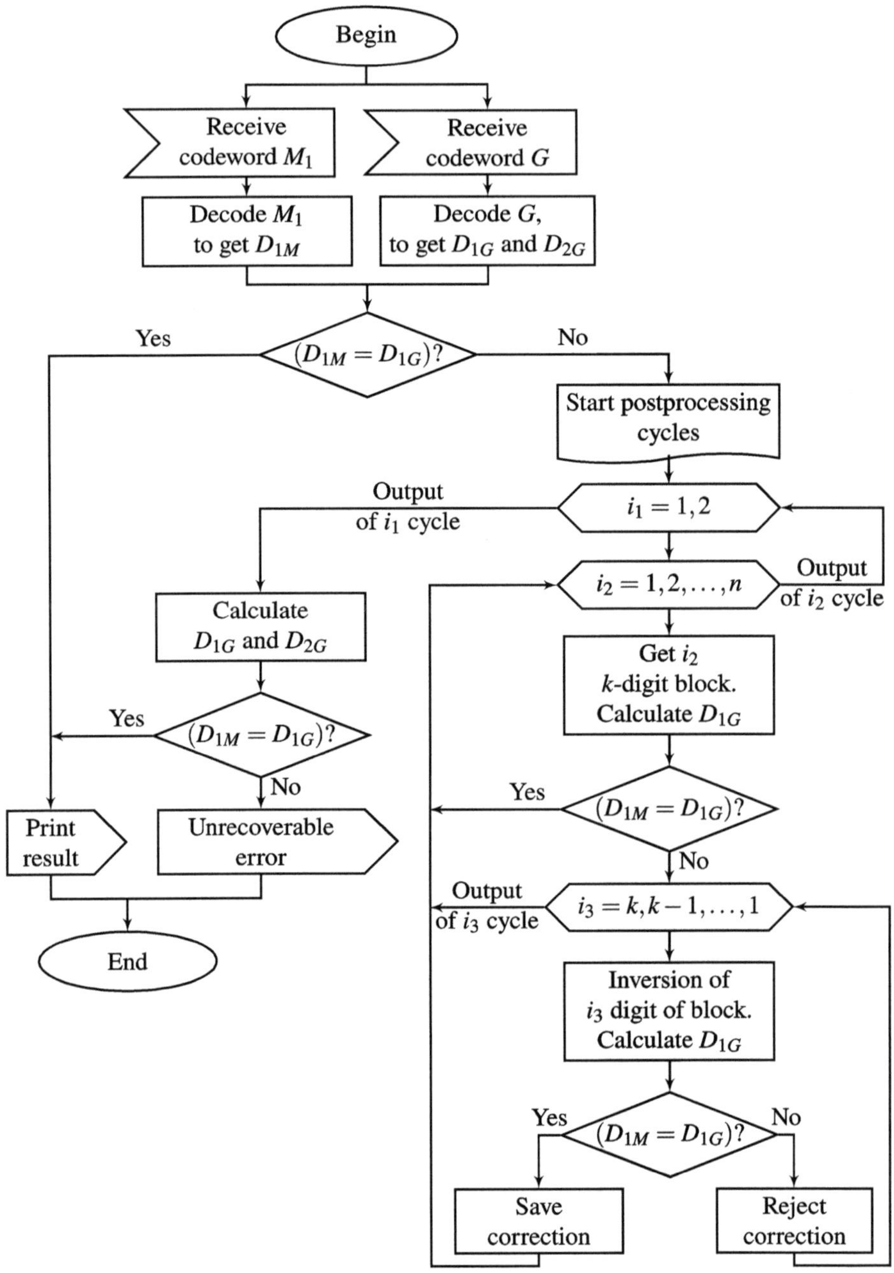

Fig. 7 Decoding of Gold code with postprocessing

The probabilistic characteristics of all considered decoding methods are shown in Fig. 8. For each of the methods, the probabilities of correct decoding, the probability

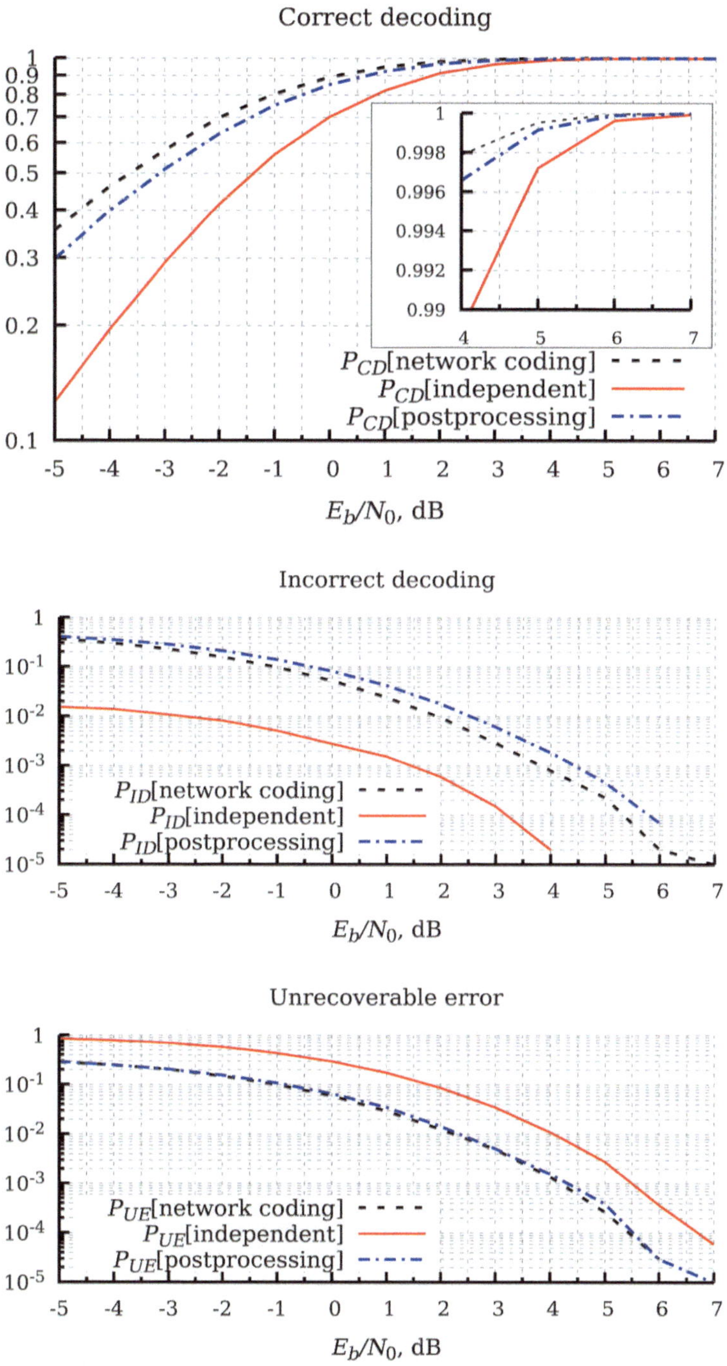

Fig. 8 Probabilistic characteristics of MLC and Gold code with network coding

of incorrect decoding, and the probabilities of detected unrecoverable error are indicated. The plots are constructed for the case of equality of the signal-to-noise ratio on the transmission routes of codewords M_1 and G (Fig. 6).

The plots use the following notation for decoding methods:

- "network coding"—the traditional method for network coding with M_2 codeword recovery and M_1 and M_2 decoding
- "independent"—independent decoding of M_1 and G codewords with comparison of decoding results
- "postprocessing"—applying postprocessing according to the algorithm in Fig. 7

5 Conclusion

Considering the obtained probabilistic characteristics, we conclude that the best result in terms of the probability of correct decoding is obtained when using the method with M_2 recovery from the predecoded word M_1. The postprocessing method shows comparable, but slightly worse results. Simple independent decoding of codewords M_1 and G has the worst probability of correct data reception of all three methods but provides a lower probability of incorrect decoding due to a larger number of detected unrecoverable errors.

In the future, it is planned to test error-correcting codes based on Kasami and Gordon–Mills–Welsh composite sequences, to evaluate their probabilistic characteristics in various network coding scenarios.

Acknowledgments The studies at St. Petersburg State University of Telecommunications prof. M.A. Bonch-Bruevich were supported by the Ministry of Science and High Education of the Russian Federation by the grant 075-15-2022-1137.

References

1. Adamson, B., Adjih, C., Bilbao, J., Firoiu, V., Fitzek, F., Ghanem, S., Lochin, E., Masucci, A., Montpetit, M.J., Pedersen, M., Peralta, G., Roca, V., Saxena, P., Sivakumar, S.: Taxonomy of coding techniques for efficient network communications. Request for Comments RFC 8406, Internet Engineering Task Force (2018). https://datatracker.ietf.org/doc/rfc8406
2. Ahlswede, R., Cai, N., Li, S.Y., Yeung, R.: Network information flow. IEEE Trans. Inf. Theory **46**(4), 1204–1216 (2000). https://doi.org/10.1109/18.850663
3. Ali, M., Kodera, Y., Kusaka, T., Uehara, S., Nogami, Y., Morelos-Zaragoza, R.: Multi-value sequence generated over sub extension field and its properties. J. Inf. Secur. **10**(3), 130–154 (2019). https://doi.org/10.4236/jis.2019.103008
4. Amanowicz, M., Krygier, J.: On applicability of network coding technique for 6lowpan-based sensor networks. Sensors **18**(6), 1–20 (2018). https://doi.org/10.3390/s18061718
5. Angelopoulos, G., Paidimarri, A., Chandrakasan, A., Medard, M.: Experimental study of the interplay of channel and network coding in low power sensor applications. In: 2013 IEEE International Conference on Communications (ICC), pp. 5126–5130. IEEE, Budapest (2013). https://doi.org/10.1109/ICC.2013.6655396

6. Ateya, A., Muthanna, A., Gudkova, I., Vybornova, A., Koucheryavy, A.: Intelligent core network for Tactile Internet system. In: Proceedings of the International Conference on Future Networks and Distributed Systems, ICFNDS '17, pp. 1–6. ACM, New York (2017). https://doi.org/10.1145/3102304.3102326
7. Ateya, A., Khayyat, M., Muthanna, A., Koucheryavy, A.: Toward Tactile internet. In: 2019 11th International Congress on Ultra Modern Telecommunications and Control Systems and Workshops (ICUMT), pp. 1–7. IEEE, Dublin (2019). https://doi.org/10.1109/ICUMT48472.2019.8970990
8. Bartz, H., Puchinger, S.: Decoding of interleaved linearized Reed-Solomon codes with applications to network coding. In: 2021 IEEE International Symposium on Information Theory (ISIT), pp. 160–165. IEEE, Melbourne (2021). https://doi.org/10.1109/ISIT45174.2021.9517834
9. Bassem, R.: Mahafza Radar Systems Analysis and Design Using MATLAB Third Edition. CRC Press, Boca Raton (2013)
10. Biczok, G., Chen, Y., Kralevska, K., Overby, H.: Combining forward error correction and network coding in bufferless networks: A case study for optical packet switching. In: 2016 IEEE 17th International Conference on High Performance Switching and Routing (HPSR), pp. 61–68. IEEE, Yokohama (2016). https://doi.org/10.1109/HPSR.2016.7525640
11. Bossert, M., Gabidulin, E.: One family of algebraic codes for network coding. In: IEEE International Symposium on Information Theory - Proceedings: 2009 IEEE International Symposium on Information Theory, ISIT 2009, pp. 2863–2866. IEEE, Seoul (2009). https://doi.org/10.1109/ISIT.2009.5205280
12. Cai, N., Yeung, R.: Network coding and error correction. In: Proceedings of the IEEE Information Theory Workshop, pp. 119–122. IEEE, Bangalore (2002). https://doi.org/10.1109/ITW.2002.1115432
13. Chen, P., Su, K., Fang, Y., Kong, L.: The design of protograph LDPC codes for channel-coded physical-layer network coding. In: 2016 IEEE 27th Annual International Symposium on Personal, Indoor, and Mobile Radio Communications (PIMRC), pp. 1–6. IEEE, Valencia (2016). https://doi.org/10.1109/PIMRC.2016.7794683
14. Feldmann, A., Gasser, O., Lichtblau, F., Pujol, E., Poese, I., Dietzel, C., Wagner, D., Wichtlhuber, M., Tapiador, J., Vallina-Rodriguez, N., Hohlfeld, O., Smaragdakis, G.: A year in lockdown: how the waves of COVID-19 impact internet traffic. Commun. ACM **64**(7), 101–108 (2021). https://doi.org/10.1145/3465212
15. Ferrett, T., Valenti, M.: Noncoherent LDPC-coded physical-layer network coding using multitone FSK. IEEE Trans. Commun. **66**(6), 2384–2395 (2018). https://doi.org/10.1109/TCOMM.2018.2801788
16. Fragouli, C., Soljanin, E.: Network coding fundamentals. Foundat. Trends Netw. **2**(1), 1–133 (2007). https://doi.org/10.1561/1300000003
17. Gabidulin, E., Bossert, M.: Algebraic codes for network coding. Probl. Inf. Transmiss. **45**(4), 343–356 (2009). https://doi.org/10.1134/S003294600904005X
18. Gabidulin, E., Pilipchuk, N.: Rank subcodes in multicomponent network coding. Probl. Inf. Transmiss. **49**(1), 40–53 (2013). https://doi.org/10.1134/S0032946013010043
19. Gold, R.: Optimal binary sequences for spread spectrum multiplexing (corresp.). IEEE Trans. Inf. Theory **13**(4), 619–621 (1967). https://doi.org/10.1109/TIT.1967.1054048
20. Halloush, R., Liu, H., Dong, L., Wu, M., Radha, H.: Hop-by-hop content distribution with network coding in multihop wireless networks. Digital Commun. Netw. **3**(1), 47–54 (2017). https://doi.org/10.1016/j.dcan.2016.08.001
21. Hammouti, M., Ar-reyouchi, E., Ghoumid, K., Lichioui, A.: Clustering analysis of wireless sensor network based on network coding with low-density parity check. Int. J. Adv. Comput. Sci. Appl. **7**(3) (2016). https://doi.org/10.14569/IJACSA.2016.070319
22. Kox, H.: Cybersecurity in the perspective of Internet traffic growth. MPRA Paper (47883) (2013). https://doi.org/10.2139/ssrn.3392430
23. Li, S.Y., Yeung, R., Cai, N.: Linear network coding. IEEE Trans. Inf. Theory **49**(2), 371–381 (2003). https://doi.org/10.1109/TIT.2002.807285

24. Liu, W., Yang, R., Pietraski, P.: Physical-layer network coding using GF(q) forward error correction codes. Research Paper, InterDigital Communications, LLC. (2011)
25. Martinez-Penas, U., Kschischang, F.: Reliable and secure multishot network coding using linearized Reed-Solomon codes. In: 2018 56th Annual Allerton Conference on Communication, Control, and Computing (Allerton), pp. 702–709. IEEE, Monticello (2018). https://doi.org/10.1109/ALLERTON.2018.8635644
26. Migabo, M., Djouani, K., Olwal, T., Kurien, A.: A survey on energy efficient network coding for multi-hop routing in wireless sensor networks. Procedia Comput. Sci. **94**, 288–294 (2016). https://doi.org/10.1016/j.procs.2016.08.043
27. Sun, Q., Yin, X., Li, Z., Long, K.: Multicast network coding and field sizes. In: 2014 IEEE International Symposium on Information Theory, pp. 2157–2161. IEEE, Honolulu (2014). https://doi.org/10.1109/ISIT.2014.6875215
28. War Hlaing, N., Farzamnia, A., Mariappan, M., Kumar Haldar, M.: Network coding schemes with efficient LDPC coded MIMO-NOMA in two-way relay networks. IET Commun. **14**(2), 337–348 (2019). https://doi.org/10.1049/iet-com.2019.0503
29. Wu, Z., Chen, X.: An optimized LDPC product network coding scheme in multiple access relay system. In: 2013 International Conference on Computing, Networking and Communications (ICNC 2013), pp. 295–299. IEEE, San Diego (2013). https://doi.org/10.1109/ICCNC.2013.6504098
30. Yastrebova, A., Kirichek, R., Koucheryavy, Y., Borodin, A., Koucheryavy, A.: Future networks 2030: Architecture & requirements. In: 2018 10th International Congress on Ultra Modern Telecommunications and Control Systems and Workshops (ICUMT), pp. 1–8. IEEE, Moscow, (2018). https://doi.org/10.1109/ICUMT.2018.8631208
31. Yeung, R., Zhang, Z.: Distributed source coding for satellite communications. IEEE Trans. Inf. Theory **45**(4), 1111–1120 (1999). https://doi.org/10.1109/18.761254

Proactive Management in Smart City: Transport Convoys

Pletnev Yaroslav

1 The Smart City Concept

The European Telecommunications Standards Institute (ETSI Technical Report) provides a definition, according to which the concept of "smart city" is a city increasing the pace and ensuring social, economic, and environmental sustainability. "A smart city must respond to challenges such as political and economic instability, climate change, rapid population growth, etc". Within the framework of the presented concept, the citizens should be provided with quality services and improve their quality of life by improving interaction with society, applying methods of joint leadership, working between different urban systems, using modern technology, and data in decision-making processes [1]. The opinion of the European Telecommunications Standards Institute in terms of the "smart city" concept was also reflected in domestic normative technical documentation. So, Technical Committee on Standardization TC 194 "Cyber-physical systems" presented PNST 443-2020 (ISO 37155-1:2020) The preliminary national standard of the Russian Federation "Information technologies. Smart City. Smart City Infrastructure." According to PNST 443-2020, the concept of "smart city" is an important tool for solving existing urban problems by integrating various city infrastructures in a rational and efficient way [2]. According to international [2] and Russian [3] organizations, the concept of a "smart city" should be based on (Fig. 1):

- Citizens' needs from the "smart city" (effective satisfaction of needs, simple and intuitive city "smart" services, transparency of information about state and municipal services, etc.)

P. Yaroslav (✉)
The Bonch-Bruevich Saint-Petersburg State University of Telecommunications, Saint Petersburg, Russia

A. A. Abd El-Latif et al. (eds.), *Secure Edge and Fog Computing Enabled AI for IoT and Smart Cities*, EAI/Springer Innovations in Communication and Computing,
https://doi.org/10.1007/978-3-031-51097-7_5

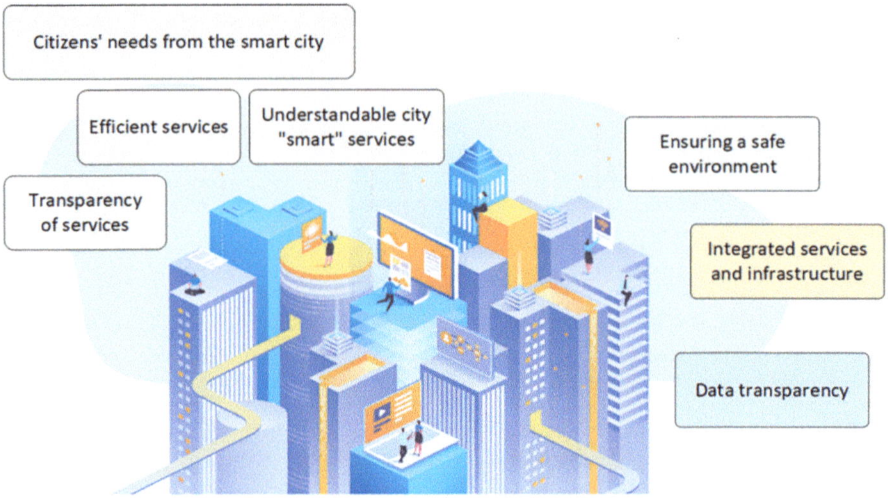

Fig. 1 Types of infrastructure projects in smart city

- Improving the performance of citizens (human centered).
- Ensuring a safe environment (including the protection of personal data and the protection of life and health).
- Has a high degree of transparency with respect to the disclosure of data on its activities.
- Is based on integrated services and infrastructure.

IBM Institute for Business Value, an organization engaged in research in the field of management and economics, providing information to business leaders on critical public and private sector issues, in its report "A vision of smarter cities – How cities can lead the way into a prosperous and sustainable future," cites a number of similar provisions [4]. One of these is that "smart cities make their systems equipped, interconnected, and intelligent". Particularly in China, IBM has helped to promote the idea that data from the city's main information systems can be integrated and interconnected through common networks and communication channels [5].

2 Consistency of Smart City Information Systems

The compatibility of urban systems components is also mentioned in the preliminary national standard of the Russian Federation "Information Technology. Smart City. Data Compatibility." It is believed that this distinctive property of smart cities can transform the capabilities of the city, ensuring informed decision-making and development of systems and services [6].

Researchers note that smart city information systems are deployed piecemeal, at the regional level, without being tied to the federal center. For example, in China, many "smart city" information systems installed in local public security departments have different technical characteristics and capabilities [5]. The heterogeneity of data from different information systems and platforms is a current problem, which has been struggled with for a long time, and it may worsen as "smart city" Internet of Things (IoT) and big data systems generate a lot of new information without a management plan [7]. It is also worth noting that data in management systems is often labeled using the language and terms of the sector in which it was collected when providing the service. Each sector uses its own models and terminologies that allow data to be discovered and understood in that sector but form a barrier to interoperability with other sectors. For example, the health sector uses the concepts of patient and treatment, the social services sector uses the concept of client, the education sector uses the concepts of learner and educational program, and the transportation sector uses the concepts of passenger and travel plan [6].

It is necessary to ensure the consistency of "smart city" information systems (Fig. 2), taking into account the goals and objectives of each of them. According to the adopted "Strategy for Digital Transformation of the Economy, Social Sphere and Public Administration in St. Petersburg," St. Petersburg is working on the development of information systems in 14 areas (figure). At the same time, during the implementation of the Strategy of Digital Transformation, the following technologies will be introduced in the city's information systems: neurotechnologies and artificial intelligence, distributed registry systems, wireless communication technologies, new production technologies, robotics and sensorics components, cloud technologies, big data collection and processing technologies, spatial analysis and modeling technologies, distributed registry system technologies, and information modeling technologies.

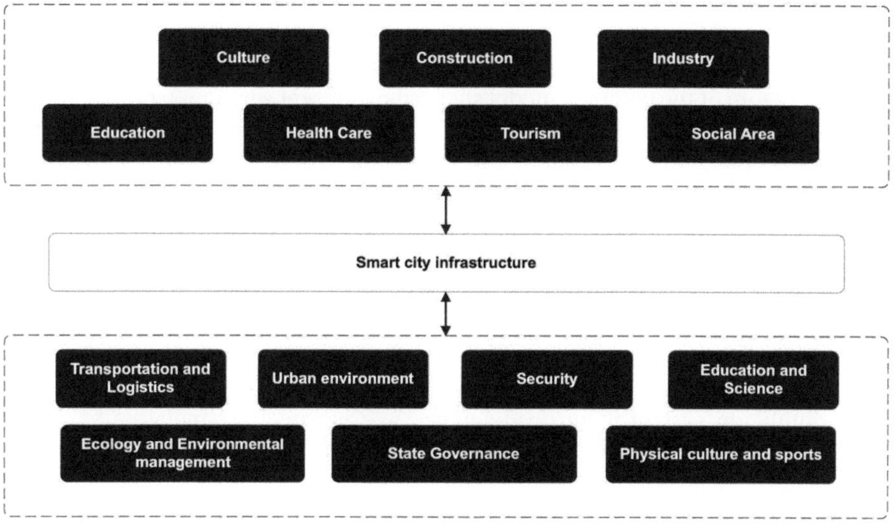

Fig. 2 Consistency of smart city information systems

A conceptual model of a smart city (SCCM) is given, which describes the overall structure of data and objects, as well as the relationships between them [8]. The SCCM proposes to work on the creation of a citywide data ecosystem that will integrate data from many sources of the city. The standard states that the reuse of data will provide better services and insights into the quality of life of city residents. One of the necessary factors is to match data from other sources with the SCCM model.

It is advisable to think through and implement the concept of data aggregator, which can achieve the purity of the data obtained, comparison with similar data from multiple information systems of the city, to ensure the interagency data exchange, to implement a clear strategy for their management.

3 System Architecture for Proactive Control of Traffic Columns in Smart City

Earlier, the author proposed a comprehensive model of the vehicle column communications modeling environment [9]. The analytical modeling and evaluation environment contains a database of analytical models, corresponding layers of communications, which are updated with the existing models tested and available for use. It is proposed to complement the proposed integrated model by adding the data aggregator of external information systems (Fig. 3).

Figure 4 shows a holistic view of the architecture of the vehicle column communication modeling system with a connected data management platform (data aggregator). In this case, the city's information systems across a variety of sectors represent subsystems of the smart city. Each such subsystem is interconnected with other subsystems and a common management platform. The platform, in turn, receives data from all subsystems, aggregates, and transmits the necessary and required information to the complex modeling environment.

The structure of the complex model shows the different simulation environments:

OMNeT++ modeling system. It is a modular object-oriented C++ discrete event network modeling system designed for modeling communication networks, multiprocessor systems, and other parallel or distributed systems; the Plexe simulation system (the Platooning Extension for Veins) is an extension of the Veins vehicle network simulator that allows you to simulate vehicle movement. It includes realistic vehicle dynamics and several cruise control models, allowing you to analyze control systems, large-scale and mixed scenarios, as well as network protocols and cooperative maneuvers;

Veins simulation system is an open-source environment for simulating vehicular networks. It is based on two simulators: OMNeT++ and SUMO, a traffic simulator.

SUMO (Simulation of Urban MObility) simulation system allows to simulate combined transportation systems, including road transport, public transport, and pedestrians.

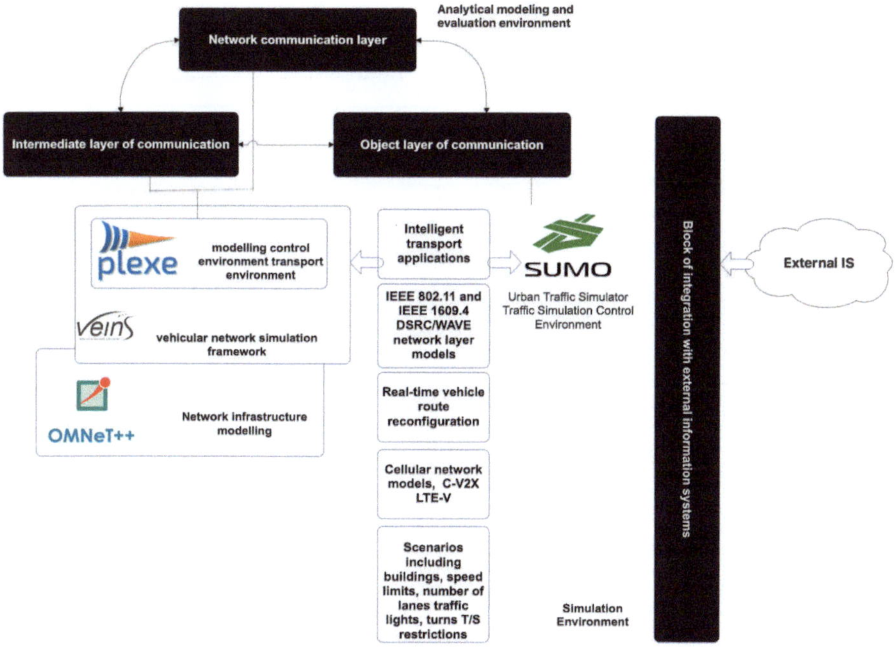

Fig. 3 Augmented comprehensive model of the vehicle column communication simulation environment

Based on the obtained data from the smart city data management platform, such as the map subsystem, it is possible to generate the required fragments of the city map for subsequent use in the simulation system (Fig. 5).

When obtaining parameters from the platform, it is also possible to simulate the situation of the passage of a convoy of vehicles, taking into account the situation on the roads at the moment. For example, simulation of situations both with low density of traffic and with high situations in areas with complex interference situations, in areas with high-rise buildings (Fig. 6).

An example of the simulation result of an intensive urban environment situation (sample of three vehicles) on artificially simulated data is shown in Fig. 7.

The use of a comprehensive model of the modeling environment in relation to columns of vehicles based on Vehicle-to-everything (V2X) technologies with the condition of inclusion of the data aggregator from the citywide data ecosystem created within the conceptual smart city model (SCCM) will allow using existing mathematical models, methods, and ways of mathematical modeling of real-time communication of transport systems. The proposed smart city data management platform (data aggregator) in this case has more comprehensive information about the state of the whole city than individual subsystems. This can be used to make more effective management decisions.

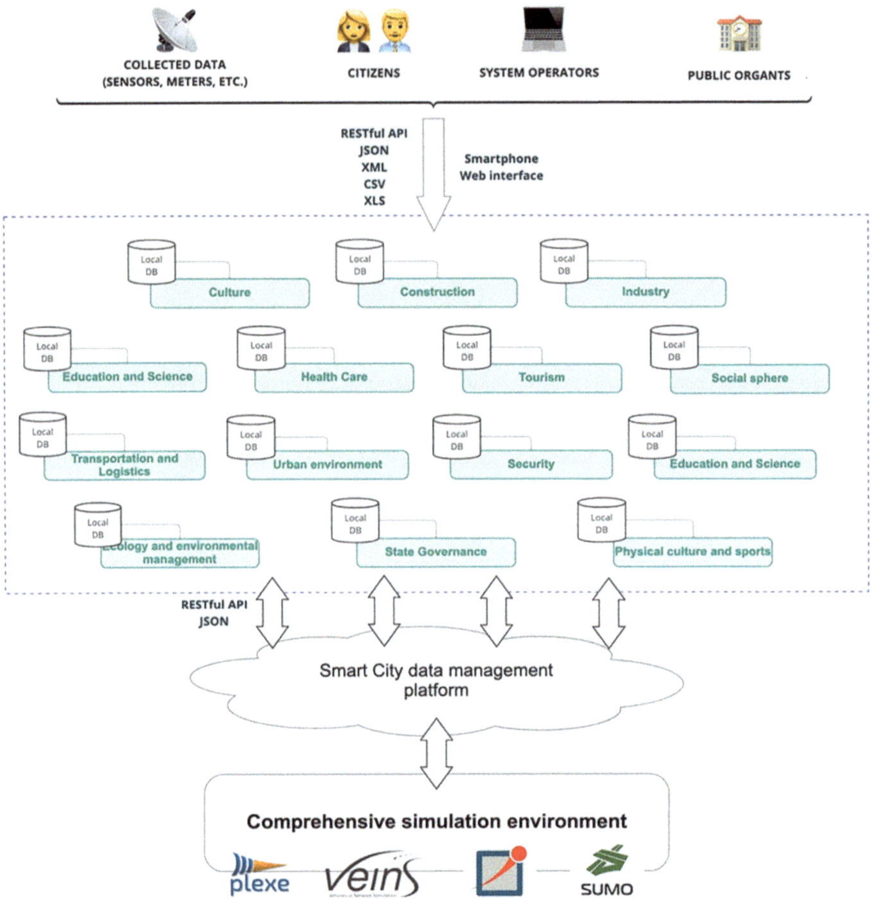

Fig. 4 Proposed architecture of the system for modeling communications of vehicle columns

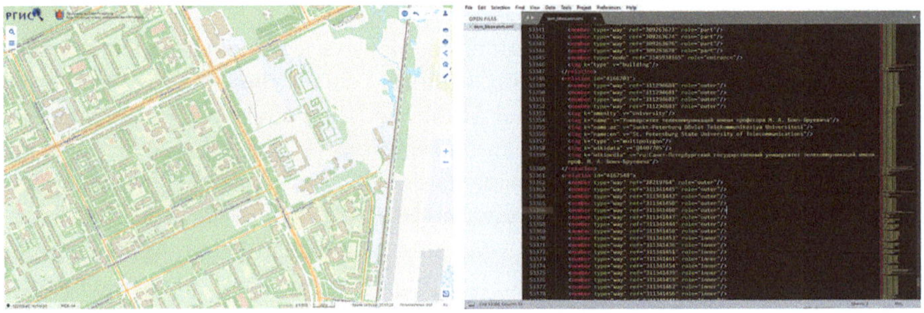

Fig. 5 Spatial data export

Fig. 6 Modeling the situation of intensive urban environment

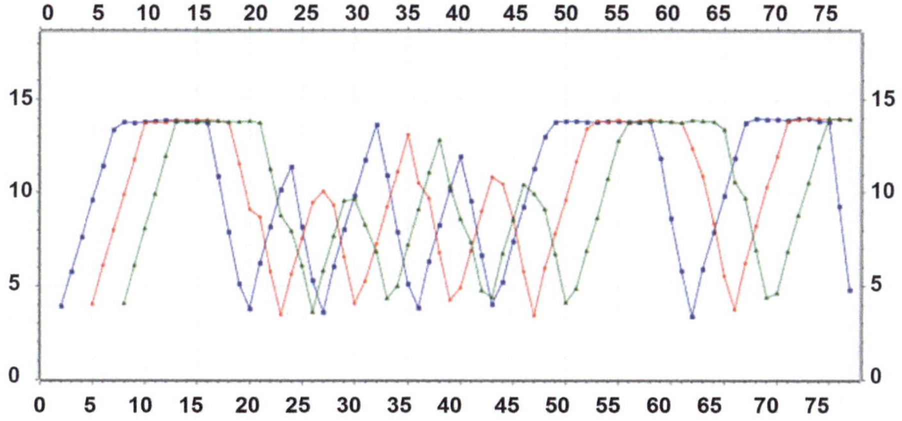

Fig. 7 Results of the simulation of the situation of intense urban environment (sample of three vehicles)

Acknowledgments The author would like to thank the editors and reviewers for their careful attention to the chapter and the comments made, which improved its quality. The chapter has been prepared in the framework of applied scientific research conducted by The Bonch-Bruevich Saint-Petersburg State University of Telecommunications in 2022.

References

1. ETSI Technical Report – DTR/HF-00 103 455 SmartCities
2. PNST 443-2020 (ISO 37155-1:2020) Information technology. Smart city. Framework for integration and operation of smart city infrastructures

3. Patentnyy landshaft «Umnyy gorod», 168 s. Proyektnyy ofis FIPS, M (2019)
4. Executive Report «A vision of smarter cities», 20s. IBM Institute for Business Value, New York (2009)
5. Research Report Prepared on Behalf of the U.S.-China Economic and Security Review Commission «China's Smart Cities Development», 119 s. SOS International LLC, Vienna (2020)
6. PNST 439–2020 (ISO 37155-1:2020) Information technology. Smart city. Data interoperability
7. Kaluarachchi, Y.: Implementing data-driven smart city applications for future cities. Smart Cities. **5**, 455–474 (2022). https://doi.org/10.3390/smartcities5020025
8. ISO/IEC 30182:2017 Smart city concept model – Guidance for establishing a model for data interoperability
9. Shestakov, A.V., Pletnev, Y.A.: «Konceptual'naya model' mnogoslojnyh kommunikacij prioritetnyh transportnyh sredstv» (Tezisy), Cifrovaya ekonomika. Novoe vremya – novye tekhnologii. ROSINFOKOM-2020. Materialy VI Vserossijskoj nauchno-tekhnicheskoj konferencii. SAMARA (2020)

Federated Learning for Linux Malware Detection: An Experimental Study

Tran Duc Le, Phuc Hao Do, Duc Tai Nguyen, Van Thang Phung, Cong Danh Nguyen, and Truong Duy Dinh

1 Introduction

Linux is an open operating system favored by programmers and engineers. Because of its advantages in terms of performance, stability, and ease of development, it is a common choice for servers and embedded devices. These systems pose an increased security risk since they might be used as a database server, network device, or control panel for a device vital for safety. It also can be attacked by different types of malware, leading to many serious consequences [1]. Modern malware [2, 3] has the capability to disguise itself and stealthily engage in various malicious activities such as stealing information, wiretapping, or penetrating systems as spyware. Additionally, it can exploit vulnerabilities in computer users' systems for extended periods. Some types focus on spreading purpose and destroying the computer system. Therefore, studying methods to detect and classify malicious codes on Linux is a very popular topic today [4].

Currently, artificial intelligence (AI) is applied in almost every field. The application of AI to detect malware is no exception [5]. The key to the success

T. D. Le (✉) · D. T. Nguyen · V. T. Phung · C. D. Nguyen
University of Science and Technology – The University of Danang, Da Nang, Vietnam
e-mail: letranduc@dut.udn.vn

P. H. Do
The Bonch-Bruevich Saint-Petersburg State University of Telecommunications, Saint-Petersburg, Russia

Danang Architecture University, Da Nang, Vietnam
e-mail: haodp@dau.edu.vn

T. D. Dinh
Posts and Telecommunications Institute of Technology, Hanoi, Vietnam
e-mail: duydt@ptit.edu.vn

A. A. Abd El-Latif et al. (eds.), *Secure Edge and Fog Computing Enabled AI for IoT and Smart Cities*, EAI/Springer Innovations in Communication and Computing,
https://doi.org/10.1007/978-3-031-51097-7_6

of AI models is data. However, a major challenge is that data security and privacy issues must be considered. The majority of currently available methods for detecting malware using machine learning (ML) or deep learning (DL) rely on a central entity to gather data from many devices and build global models. These models are then deployed to specific clients, or these clients provide the server with real test data for malware and behavior analysis. This strategy is ineffective, nevertheless, in situations when device actions involve sensitive or private information. A similar circumstance arises when human people and/or private acts are engaged in the monitored data sources.

Federated Learning (FL) [6–8] is considered an approach to overcoming the challenges of data sensibility. Federated learning refers to a configuration in which several clients work together to solve machine learning issues while managed by a central aggregator. Additionally, this configuration permits decentralized training data, which protects each device's data privacy [9]. Local computation and model transmission are the two main aspects of federated learning, which lowers some systemic costs and privacy problems associated with centralized techniques. The client's original data is kept locally and cannot be swapped or changed. To accomplish the learning objective, federated learning is applied. Each device utilizes local data for local training and uploads the model to the server for aggregation. Then the server delivers the participants the updated model. A central entity known as a server can share model parameters and aggregate them to generate a singular and global model [6]. Each client has a global model after a number of iterations that were created by combining each client's models. Since data is not shared with external identities, this strategy naturally supports data privacy.

Malware on Linux often takes the form of ELF (Executable and Linking Format) executable files. Every ELF file has three key sections: *ELF Header, Program Header Table, and Section Header Table* (Fig. 1). An executable or shared object file describes a segment or other information the system needs to prepare the program for execution in the program header table, which comes after an ELF header that explains the structure of a file. An executable code is always placed in the *.text (code) section*, while the *.rodata section* provides read-only data. Each section

Fig. 1 ELF file format

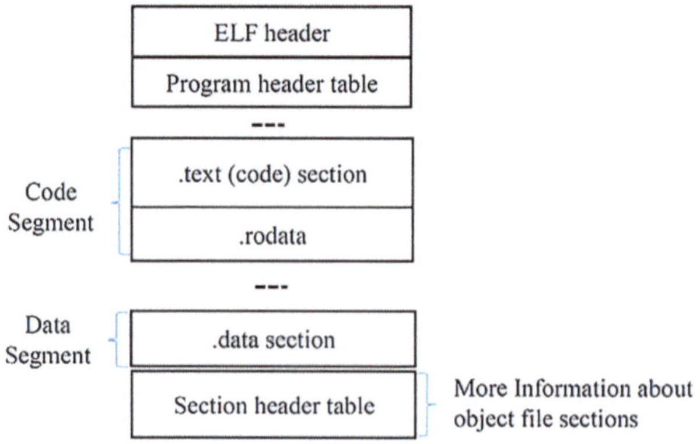

represents a portion of a file. All user-initialized data variables are put in a *.data section* of the data segment. A section header table provides additional details about a file's sections.

The research questions are: (1) How to apply a federated learning approach to detect malware in Linux? (2) Can the federated learning model replace the traditional centralized approach? (3) How to analyze and extract essential features from malware ELF files to increase the model's efficiency?

In this study, we aim to answer those research questions. We will compare the result of the FL approach with the traditional approach to see the possibility of the FL model in detecting Linux malware. In addition, to improve the performance of the FL model, we focus on using essential features of the ELF format to detect Linux malware.

The remainder of this paper is organized as follows: in Sect. 2, related works are discussed. Section 3 introduces the framework and deployment for detecting malware on Linux. Section 4 expresses our implementation plan, experiments, and results. Finally, Sect. 5 presents conclusions and future work.

2 Related Works

Many studies have been conducted on identifying malware for the Windows-based operating system [10–12]. However, finding malware for Linux-based operating systems is rare.

According to Asmitha, K.A. and P. Vinod in [13], a machine learning-based strategy that employs system calls as features were presented. System calls are strong characteristics because they may show how a process behaves. The model dynamically extracts system calls from Linux executables, and the best feature set with the highest potential for categorization is chosen. The author's objective is to categorize both malicious and safe Linux executables.

Igor Vurdelja et al. in [14] provided a summary of methods for detecting malware on Linux systems using system tracers. Different ways of malware detection are explored. The authors noted that while most techniques demonstrated good classification accuracy, they could be deficient in relevant datasets and true predictive potential. Using system tracers to identify Linux malware requires a larger, better-prepared machine learning dataset and a realistic execution simulation, according to a critical examination of the linked research.

In [15], Asmitha, K.A., and Vinod, P. introduced a non-parametric statistical method utilizing machine learning approaches for detecting harmful Executable Linkable Files. System calls are used as dynamically extracted characteristics in a controlled environment. The suggested method ranks and identifies key attributes using non-parametric statistical techniques such as the Kruskal–Wallis ranking test (KW), Deviation From Poisson (DFP), and others. A basic collection of characteristics taken from the system called traces is used to construct a prediction model, which is then applied by three learning algorithms (J48, AdaBoost, and

Random Forest). It can be deployed in real-time since both model development and prediction have a low overhead.

Up to now, several studies have applied federated learning in the field of malware analysis and detection. In research [16], Kuang-Yao Lin and Wei-Ren Huang proposed a prototype for classifying malware that used a decentralized data-collecting mechanism. The authors compared the impacts of SVM, conventional learning, and federated learning in experiments that simulated training circumstances. On the dataset supplied by VirusTotal, the results showed a remarkably high degree of accuracy.

In [17], Valerian Rey et al. described a method for detecting malware that affects IoT devices using federated learning. This framework is made to be installed on network nodes that give IoT devices WiFi and 5G access. The authors compared the performance of three different setups: a local setup where each device owner trains a single, isolated model; a federated approach where each device owner trains their model, which is periodically aggregated in a server; and a non-privacy-preserving setup where the entire dataset is centralized and trained by the server. This study has demonstrated that using bigger, more varied datasets–like those used in federated and centralized methods–significantly influenced the model's performance in both supervised and unsupervised settings. Additionally, it has been shown that using a federated strategy may maintain data privacy without degrading model performance. The positive results from studies [16, 17] prompted us to experiment with federated learning in a new environment—Linux operating system.

3 Federated Learning Approach for Linux Malware Detection

This section describes the architectural design of the proposed FL-based framework in depth. We also go through each of its elements and how they relate to one another during model training and evaluation. In Fig. 2, the framework architecture comprises k clients, each with its dataset. The server manages the FL procedure. The design information for each part of the proposed architecture is provided in the subsequent sections.

3.1 Data Collection

This study uses sample data generated on each client, including malware or benign files in ELF format. The malware dataset is manually collected from many sources,

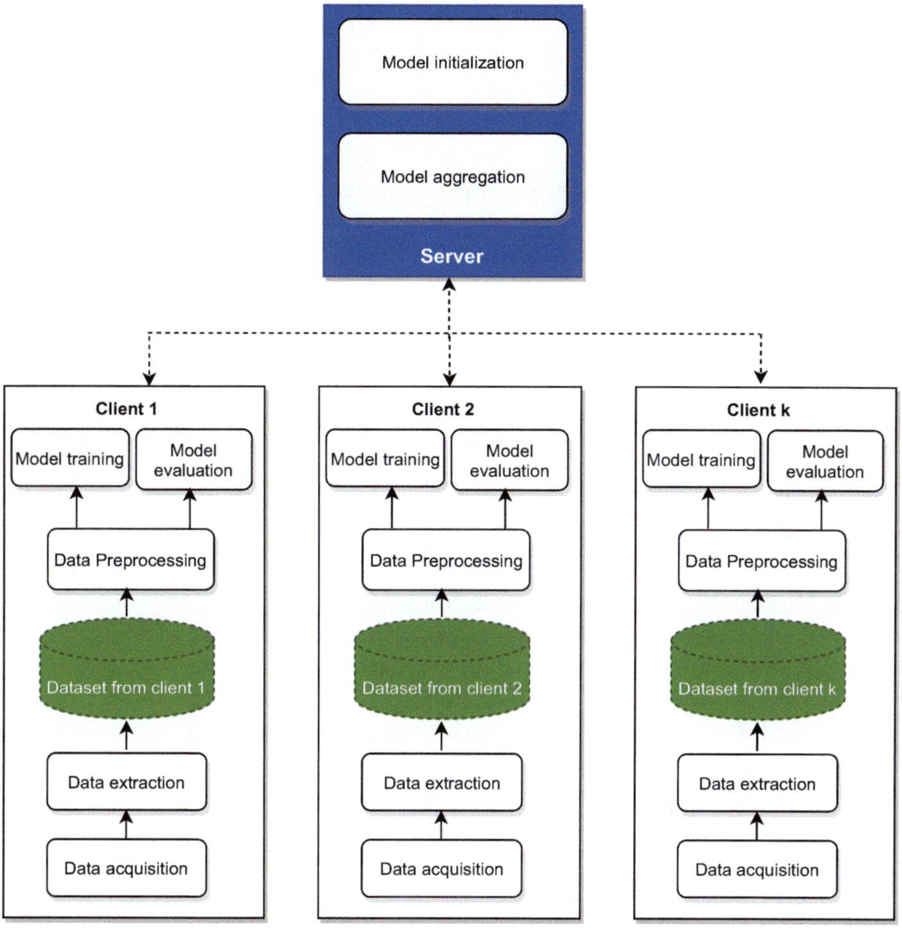

Fig. 2 Framework architecture and its components

such as *Virusshare*,[1] *VirusTotal*,[2] and *GitHub*.[3] However, because these malware samples are of many types, we only select 32-bit ELF files to have a unified dataset. Using the *pyelftools-master*[4] library, we analyzed and collected 2900 samples, half benign files, and half were Linux malware.

[1] https://virusshare.com/,

[2] https://www.virustotal.com/.

[3] https://github.com/.

[4] https://github.com/eliben/pyelftools.

Table 1 Fields in ELF header

Field	Meaning
e_ident[];	/*Magic number and other info */
e_type;	/* Object file type */
e_machine;	/* Architecture */
e_version;	/* Object file version */
e_entry;	/* Entry point virtual address */
e_phoff;	/* Program header table file offset */
e_shoff;	/* Section header table file offset */
e_flags;	/* Processor-specific flags */
e_ehsize;	/* ELF header size in bytes */
e_phentsize;	/* Program header table entry size */
e_phnum;	/* Program header table entry count */
e_shentsize;	/* Section header table entry size */
e_shnum;	/* Section header table entry count */
e_shstrndx;	/* Section header string table index */

Since most of the malware's features are concentrated in the *ELF Header* and Section Header, we mainly extract data from these two components in this study. The structure of an *ELF Header* will include the fields presented in Table 1.

Details of the workflows for the entire data preprocessing, training, and evaluation process are shown in Fig. 3. For simplicity, the server's first model sharing is skipped. Before the model is assessed, steps 1, 2, 3, and 4 are supposed to be performed multiple times (step 5)

3.2 Data Preprocessing

During the data preprocessing, we use several techniques to make the data cleaner and better to increase the accurate detection rate. One of these techniques is called min-max scaling (often called "normalization"). It is a different method from Z-score normalization (or standardization). This method scales the data to a defined range, often from 0 to 1. In contrast to standardization, the cost of having this bounded range is that we will end up with smaller standard deviations, which can suppress the effect of outliers.

$$X_{norm} = \frac{X - X_{min}}{X_{max} - X_{min}} \tag{1}$$

where X_{norm} is the normalized value

In addition, the dataset includes multiple columns with numerical and categorical variables, but the model can only comprehend numbers. It is unable to interpret the text. Therefore, categorical columns must be transformed into numerical columns for the model to understand them. We use a one-hot encoding technique to convert categorical data into numerical data to normalize the original data. After the data preprocessing, we will have enough clean data as input for the system's detection process.

3.3 Training Model

The training model parameters of each kth client will have the corresponding weight w^k. Suppose the global model parameters are w. In the training process, the model will find the optimal weight w^k. The w^k weights will be sent directly to the server. The server will get the optimal weights of each client, and then it will use the average value of the weights, that is:

$$w = \frac{1}{k}\sum_{1}^{k} w^k \tag{2}$$

The new aggregated global model is then distributed back to the Model Training component of the clients to continue the training process. This process is repeated until a certain number E of epochs spanning the whole train set have been accomplished. It should be noted that in this experiment, we use the average of the weights at the server. In practice, however, the weights may vary, and using the mean may not be the most efficient. Here we can perform the optimization problem for the weights. This problem is beyond the scope of the study.

3.4 Model Evaluation

After the model has been trained with a certain number of iterations according to the flow in Fig. 3, we can evaluate the model. To evaluate the model, we use the optimized weights sent from the server and conduct tests on each test dataset of each respective client.

3.5 Server

The server receives data, synthesizes it, and sends the necessary parameters to the clients. The server will first initialize the default weight for all models at clients. It means that the clients will have the same initialization weight at initialization.

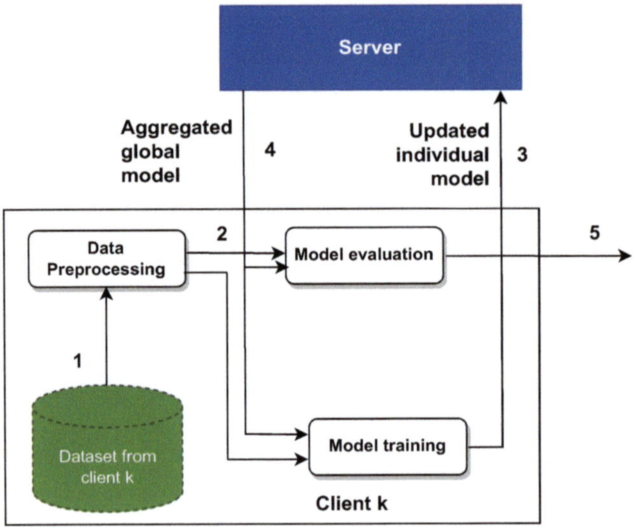

Fig. 3 The client components and workflows

On the client side, based on the initial training data with the weights received from the server, the client will train the model and calculate the new weight values for that particular client. The new weight is then sent back to the server.

After receiving new weights from different clients, the server will aggregate the parameters. This study uses the average method to update the common weights for all clients.

4 Experiment

The experiment was conducted following the procedure described in Sect. 3. We implemented two different scenarios to compare the effectiveness of Federated Learning and Fully Connected Models: (1) Using a centralized dataset to train a Fully Connected Model at one server. (2) Using a central server and four clients for training. The dataset is collected at each client in the form of ELF files. The dataset consists of 2900 files, with 50% of the software being malware and the rest benign.

The evaluation was performed on a Windows 10, 64-bit computer with Core AMD A8-4555M APU (1.6 GHz, 4 cores, 4 processors), and 8 GB RAM. Due to static analysis, the computation cost is very low. Hence, all experiments are carried out on the typical end-host system.

4.1 Training Process

Initially, the central server is installed, initializing a training model. The server sends copies of that model to the clients. Then the model instance will run in each of the clients.

Once the training is complete on the local data, the client connects to the server, and the model weights/parameters are sent to the central server. The central server aggregates the weight parameters (by averaging) and sends the updated weights back to each client. Thus ensuring that data never leaves the client and we get the benefit of training on the data from all the clients. This process can occur continuously and repeatedly until the model achieves high accuracy. Once the clients receive the latest updated parameters, they can use the updated model parameters for their tasks.

In this experiment, we do 1000 training times for the traditional model. For the federated learning model, we perform ten iterations, and in each such iteration, each client will perform 100 training times.

4.2 Results

The results obtained from training using a traditional fully connected model with a centralized dataset at the server and a batch size of 64 over 1000 epochs are illustrated in Fig. 4. The value of the loss function is low. It represents a very good convergence model on the processed dataset. The results are train_loss = 0.0736 and val_loss = 0.1424 (validation). The accuracy of the model has been converged. The corresponding results obtained on the train set and the test set are train_accuracy = 0.9608 and test_accuracy = 0.9259.

The results obtained after training using Federated Learning are shown in Fig. 5 with distributed dataset and batch_zize = 64, epochs = 100, the number of clients $k = 4$, is also very positive (train_loss = 0.1, val_loss = 0.2, train_accuracy = 0.87, test_accuracy = 0.87).

As can be seen, the results obtained when training with the traditional model is higher than that of the FL model. It is understandable since the traditional model performs training with a centralized dataset. The difference in weights will not be affected as in the FL model. However, we also find that the FL model gives good results even though the data is distributed across different clients. This result is achieved because the model is evaluated and trained many times through loops. Then the weights will be better chosen and closer to the optimal value on the client side. It should be noted that since the present study only aggregates the weight parameters by averaging at the server, the results are not optimal.

Besides, we also found that the time to train the traditional model is usually shorter than that of the FL model. The reason is that the data of the traditional model is concentrated on the server when training, while the FL model needs additional

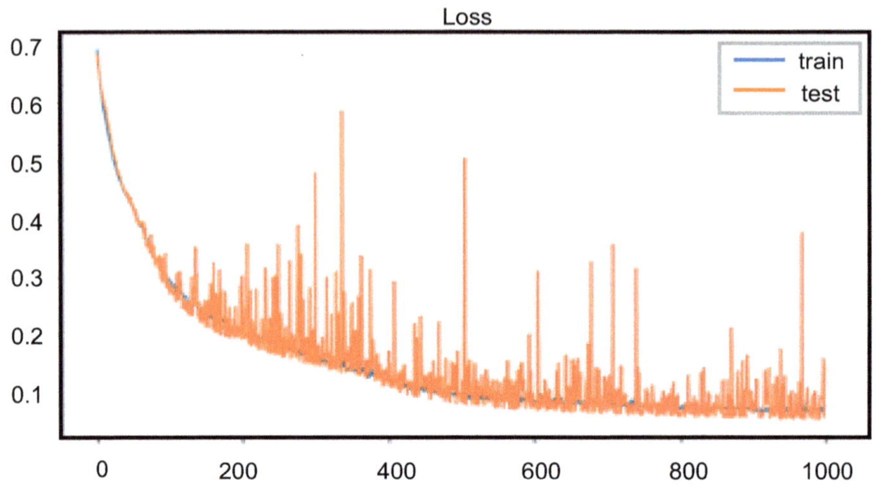

(a) Loss function graph with the train set and test set

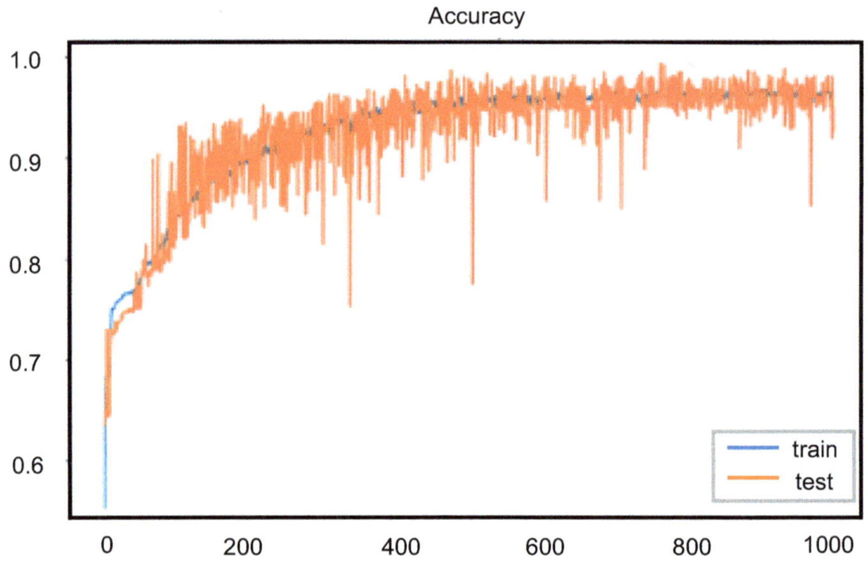

(b) Graph of accuracy with the train set and test set

Fig. 4 Results obtained when using the traditional fully connected model with centralized data. (**a**) Loss function graph with the train set and test set. (**b**) Graph of accuracy with the train set and test set

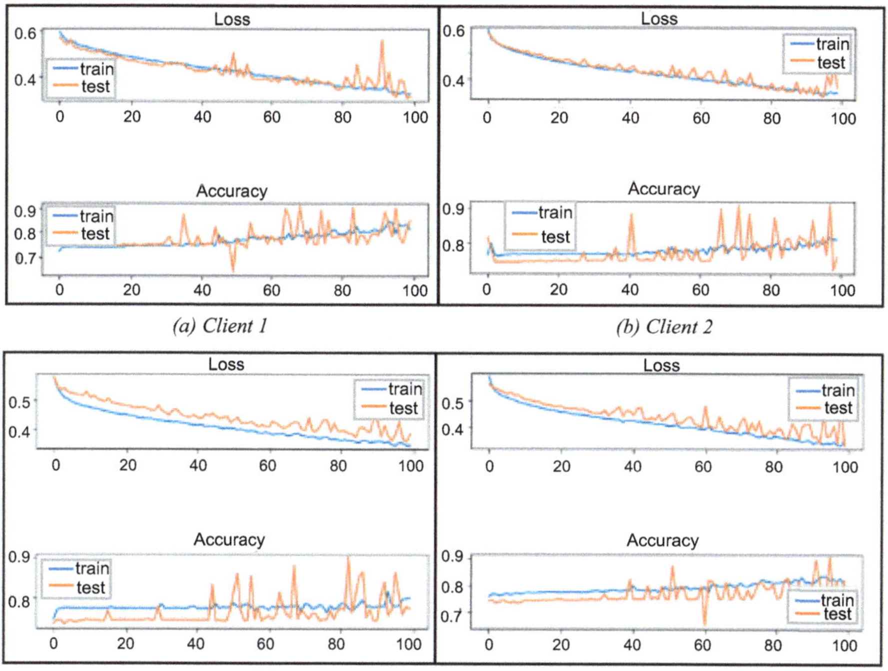

(a) Client 1 (b) Client 2 (c) Client 3 (d) Client 4

Fig. 5 Results obtained when using federated learning model with distributed data. (**a**) Client 1. (**b**) Client 2. (**c**) Client 3. (**d**) Client 4

steps such as sending, receiving, calculating, and aggregating the weights. But obviously, the FL model has a much higher level of data security and privacy than the traditional model because clients do not need to send data to anyone.

5 Conclusions

Data privacy and security during training machine learning or deep learning models today is of great concern. To overcome this challenge, the training data must be decentralized. The Federated Learning model was born from that need. However, a question arises with such decentralized data, how effective will the training models be? Can the FL model replace the traditional model with centralized data on a server?

In this study, we experiment with the FL model for detecting malware in the Linux operating system. We experimentally want to compare the traditional Fully Connected Models with centralized data and the FL model with distributed data.

Although the FL model has not been optimized for parameters and weights, the results obtained are quite positive compared to the traditional model. This result can also be improved if the weights are optimized from the client side and if there is an appropriate way to aggregate the best weights at the server instead of calculating the average value.

In the future, our goal is to expand the number of ELF files and ELF 64-bit files in the dataset and extract more of the Program Headers of those files. At the same time, improve another more complex model. Moreover, we will evaluate the accuracy of the clients after each training and perform model synthesis at the central server according to the accuracy ratio.

References

1. Mohanta, A., Saldanha, A.: Persistence Mechanisms. In: Malware Analysis and Detection Engineering, pp. 213–236. Apress, Berkeley (2020)
2. Andrea, I., Chrysostomou, C., Hadjichristofi, G.: Internet of things: Security vulnerabilities and challenges. In: 2015 IEEE Symposium on Computers and Communication (ISCC), pp. 180–187, 2015. Afianian, A., Niksefat, S., Sadeghiyan, B., Baptiste, D.: Malware dynamic analysis evasion techniques: a survey. ACM Comput. Surveys **52**(6), 1–28 (2019)
3. Alaeiyan, M., Parsa, S., Conti, M.: Analysis and classification of context-based malware behavior. Comput. Commun. **136**, 76–90 (2019)
4. McMahan, B., Moore, E., Ramage, D., Hampson, S., Arcas, B.A.: Communication efficient learning of deep networks from decentralized data. In: Singh, A., Zhu, J. (Eds.) Artificial Intelligence and Statistics. In: Proceedings of Machine LearningResearch, vol. 54, pp. 1273–1282. PMLR, Fort Lauderdale (2017)
5. Yang, Q., Liu, Y., Chen, T., Tong, Y.: Federated machine learning: concept and applications. ACM Trans. Intell. Syst. Technol. **10**(2), 1–19 (2019)
6. Yang, Q., Liu, Y., Cheng, Y., Kang, Y., Chen, T., Yu, H.: Federated learning. Synth. Lect. Artif. Intell. Mach. Learn. **13**(3), 1–207 (2019)
7. Joshi, S., Upadhyay, H., Lagos, L., Akkipeddi, N.S., Guerra, V.: Machine learning approach for malware detection using random forest classifier on process list data structure. In: Proceedings of the 2nd International Conference on Information System and Data Mining, pp. 98–102 (2018)
8. Kumar, N., Handa, A.: Detection of advanced linux malware using machine learning. In: Advances in Computational Intelligence and Communication Technology, pp. 185–194. Springer, Singapore (2021)
9. Kairouz, P., McMahan, H.B., et al.: Advances and open problems in federated learning. Found. Trends®Mach. Learn. **14**(1–2), 1–210 (2021)
10. Lyu, L., Yu, H., Ma, X., Sun, L., Zhao, J., Yang, Q., Yu, P.S.: Privacy and robustness in federated learning: Attacks and defenses (2020). Preprint arXiv:2012.06337
11. Biggio, B., Nelson, B., Laskov, P.: Poisoning attacks against support vector machines. In: Proceedings of the 29th International Coference on International Conference on Machine Learning, pp. 1467–1474 (2012)
12. Blanchard, P., El Mhamdi, E.M., Guerraoui, R., Stainer, J.: Machine learning with adversaries: Byzantine tolerant gradient descent. In: Proceedings of the 31st International Conference on Neural Information Processing Systems, pp. 118–128 (2017)
13. Yin, D., Chen, Y., Kannan, R., Bartlett, P.: Byzantine-robust distributed learning: Towards optimal statistical rates. In: International Conference on Machine Learning. PMLR, pp. 5650–5659 (2018)

14. He, L., Karimireddy, S.P., Jaggi, M.: Byzantine-robust learning on heterogeneous datasets via resampling (2020). Preprint arXiv:2006.09365
15. Christoforou, E., Anta, A.F., Georgiou, C., Mosteiro, M.A., Sánchez, A.: Applying the dynamics of evolution to achieve reliability in master–worker computing. Concurr. Comput. Pract. Exper. **25**(17), 2363–2380 (2013)
16. Haddadpour, F., Mahdavi, M.: On the convergence of local descent methods in federated learning (2019). Preprint arXiv:1910.14425
17. Yang, Q., Liu, Y., Chen, T., Tong, Y.: Federated machine learning: Concept and applications. ACM Trans. Intell. Syst. Technol. **10**, 1–19 (2019)

Delay Prediction in M2M Networks Using the Deep Learning Approach

Ali R. Abdellah, Malik Alsweity, Mohamed H. Essai, Ammar Muthanna, and Andrey Koucheryavy

1 Introduction

The main feature of the fifth-generation (5G) networks is the development of the Internet of Things (IoT) concept. This feature is associated with a fundamentally different number of potential things that can be connected to the communications network and uniquely identifiable within it, in contrast to conventional ideas about the size of device databases on the network. According to one well-known prediction, the limit for the Internet of Things will be 50 trillion things. Due to the large number of connected devices, each unit area has a high density [1].

The Internet of Things (IoT) is a network of interconnected, sensor-equipped, communicating devices that enable automatic data collection and sharing. Machine-to-machine (M2M) communication is a type of communication in which devices can communicate. Cellular networks allow communication over both short and long distances. Data collection, M2M communication, and application processing are the three primary functions of IoT devices.

Infrastructure must use techniques to manage, store, and analyze large amounts of data because IoT devices are predicted to handle enormous amounts of data. IoT devices can provide services such as monitoring, managing nodes, storing, and analyzing data. IoT devices are primarily utilized in applications for healthcare, home appliances, transportation, environmental monitoring, logistics management,

A. R. Abdellah · M. H. Essai
Electrical Engineering Department, Faculty of Engineering, Al-Azhar University, Qena, Egypt
e-mail: alirefaee@azhar.edu.eg; mhessai@azhar.edu.eg

M. Alsweity · A. Muthanna (✉) · A. Koucheryavy
Department of Telecommunication Networks and Data Transmission, The Bonch-Bruevich Saint-Petersburg State University of Telecommunications, St. Petersburg, Russia
e-mail: muthanna.asa@spbgut.ru

A. A. Abd El-Latif et al. (eds.), *Secure Edge and Fog Computing Enabled AI for IoT and Smart Cities*, EAI/Springer Innovations in Communication and Computing,
https://doi.org/10.1007/978-3-031-51097-7_7

security services, and a wide range of other applications. The effectiveness of an IoT device is primarily determined by how easily it can communicate with other IoT devices. Its efficiency is assessed. Increasing speed reduces delays and enhances data transfer between devices. Consistent and stable network conditions are critical for any IoT system that relies on real-time updates. The network's ability to handle more connected devices increases reliability [2].

The most recent wireless mobile technology is the 5G network. There is enormous network traffic due to the increasing proliferation of network applications. Numerous researchers have focused on applications that accurately predict network traffic. The ability to predict network traffic without the involvement of the network operator is required for network management. To operate in high-traffic networks where multiple sensors send their packets to a destination simultaneously, a feature of IoT applications, 5G in practice, leverages the Internet of Things (IoT). As a result, 5G offers very high data throughput, low delay, and high bandwidth. Network traffic prediction is even more critical for IoT networks to provide reliable communications. For a 5G network to work effectively, artificial intelligence (AI) methods must be used. Machine learning (ML) to predict traffic has already been successful [3, 4].

With the advent of big data–enabled data and the invention of high-speed network components, Internet traffic has increased tremendously. The resources of such a network are indispensable and must be used for their intended purpose. Moreover, it is a challenging task to monitor and predict traffic. There are many artificial intelligence-based prediction techniques for predicting traffic. Some methods and techniques have already been applied, such as predicting daily Internet traffic using a data mining technique, a low complexity–based machine learning boost, a classification and regression algorithm for predicting Internet traffic from weak to strong learning, a double exponential predictor based on artificial neural networks (ANN), classical time series, and wavelet transform–based predictors, a deep learning (DL)-based prediction for traffic in metropolitan areas, and a neural network ensemble for predicting Internet traffic [3, 4].

This study uses deep learning (DL) and a long short-term memory (LSTM) network to predict delays in M2M networks. We use a time-series prediction method to predict M2M traffic. Root mean square error (RMSE) and mean absolute percentage of error (MAPE) are used to evaluate the prediction accuracy. To investigate the effect of changing the number of epochs used in training the DL model on maximizing prediction accuracy, we proposed three DL models based on the number of epochs used in training the DL model.

The outline of this chapter is as follows: Sects. 1: Introduction; 2: Related work; 3: Problem statement and system modeling; 4: Experimental results; and 5: Conclusion.

2 Literature Review

Various researchers have become interested in using ML algorithms to predict wireless network traffic in 5G technologies. In this research, we address the delay prediction of IoT and tactile Internet traffic using ML, especially the DL prediction strategy with an LSTM network. Therefore, this part includes the earlier work relevant to our research area.

The time-series prediction technique for IoT traffic was created with Managed Service Provider (MSP) using Nonlinear Autoregressive with Exogenous Inputs (NARX)- recurrent neural network (RNN), and the prediction accuracy in terms of the loss functions Mean Squared Error (MSE), Sum of Squared Errors (SSE), Mean Absolute Error (MAE), and MAPE was evaluated in [5]. Using DL, a multiparametric technique, Ateeq et al. [6] developed an algorithm to predict delays in IoT traffic. In smart grid applications, an LSTM-ANN is used to estimate the delay in the communication network [7]. Time-series forecasts for IoT and tactile Internet delays are produced by Abdellah et al. [8] using a k-step-ahead prediction approach and an RNN with NARX support. Predicting IoT's SNR and packet delivery ratio (PDR) is performed using deep learning–enabled multiparametric forecasting [9]. An LSTM network has been used to study the prediction of network traffic based on DL [10–13]. Using ML regression techniques, Hameed et al. [14] predicted throughput in an IoT network. The ability of DL to forecast traffic flow in autonomous vehicle cellular networks has been investigated by Ali Shah et al. [15]. Time-series methods with statistical learning were used to predict traffic produced by IoT systems [16]. To forecast traffic in V2X networks, DL based on the unidirectional LSTM model is proposed [17].

3 Problem Statement and System Modeling

Nowadays, several ML techniques, including random forest [18], ARIMA [19], NARX [5, 8, 20–22], and others, are used to predict network traffic. This chapter focuses on time-series prediction for M2M traffic using DL and an LSTM model. We first created the IoT system for dataset generation. We used the AnyLogic simulator to model the IoT system—the collected and prepared dataset as input to ANN to generate the prediction. After loading the dataset into the network as input and splitting it into two subsets in the input (I) and output (O) columns, it is divided into training and testing subsets. The dataset was normalized to be between [0, 1]. DL was trained in different scenarios depending on the number of training epochs used. In addition, we calculated the RMSE and MAPE for prediction accuracy.

A. *IoT System Model.*

This section presents the IoT model used to create the dataset for ML training shown in Fig. 1. The IoT system was developed using AnyLogic Simulation, which supports discrete event simulations.

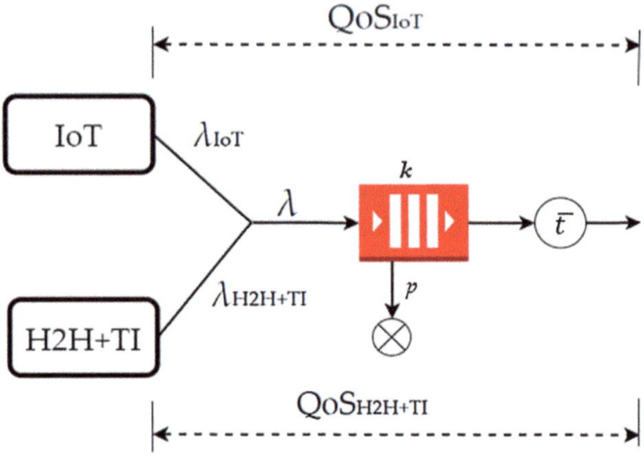

Fig. 1 Model of the IoT system

Figure 1 depicts the structure of an IoT model to simulate the process of one or more IoT devices. The traffic generator consists of traditional communication services and tactile Internet (TI) traffic implemented as H2H+ TI (H2H – human to human, TI – tactile Internet). Incoming traffic is routed to the node connection, and the model is implemented as a queuing system with integrated service discipline (with a delay and failure basis system). The average processing time of a packet (message) is equal to $\overline{t}\times$.

The arrival rate of IoT traffic is denoted by λ_{IoT}, that of H2H traffic by λ_{H2H}, and the combined flow is $\lambda = \lambda_{\mathrm{H2H}} + \lambda_{\mathrm{IoT}}$. With probability p, a packet arrives at the system input with all positions in the queue occupied, and a denial of service occurs (losses occur). The combined traffic flow at the system output has an overall intensity of λ. The characteristics of both flows define the collected traffic flow at the system input, which generally differs from the characteristics of conventional traffic and traffic in the IoT.

The queuing system (QS) model can be implemented as a G/G/1/k system. There are no accurate mathematical methods for evaluating the chance of packet loss and delivery delay in this system (waiting time in the queue). The propagation approximation approach is used [20, 21] to predict the likelihood of packet loss given known distribution parameters defining the incoming traffic and the packet processing process. There are no accurate analytical methods for evaluating the probability of packet loss and delivery delays in this system (waiting time in the queue).

The diffusion approximation approach is utilized to predict the probability of packet loss given known distribution parameters defining the inbound traffic and packet service process. The following expression is produced as an approximation:

$$p = \frac{1-\rho}{1-\rho^{\frac{2}{C_a^2+C_s^2}n_b+1}}\rho^{\frac{2}{C_a^2+C_s^2}n_b} \tag{1}$$

where $C_a{}^2$ and $C_s{}^2$ are the quadratic distribution coefficients of variation for incoming data flow and service time, respectively, nb is the buffer size, and s is the system load. To create an estimate of the average packet delivery time, use the following formula:

$$T = \frac{\rho \bar{t}}{2(1-\rho)} \left(\frac{\sigma_a^2 + \sigma_s^2}{\bar{t}^2} \right) \left(\frac{\bar{t}^2 + \sigma_s^2}{\bar{a}^2 + \sigma_s^2} \right) + \bar{t} \tag{2}$$

B. *Implementation of RNN and LSTM*

The idea of RNN is to use information consistently. In traditional neural networks, all inputs and outputs are assumed to be independent. However, this is not appropriate for many tasks. If one wants to predict the next word in a sentence, it is better to consider the preceding words. RNNs are called recursive because they perform the same task for each sequence element, with the output depending on the previous computations. Another interpretation of RNNs is that they have a "memory" that considers previous information. RNNs can theoretically use information in sequences of any length, but in practice, they are limited to a few steps. Figure 2 shows the schematic of a single-layer recurrent neural network. At each cycle, the inner layer of neurons receives a set of input data X and information about the previous state of the inner layer A, based on which it generates a response h.

Mathematically, a simple RNN can be formulated as follows:

$$h(t) = f_h \left(w_{ih}\, x(t) + w_{hh}\, h\,(t-1) \right) \tag{3}$$

$$y(t) = f_o \left(w_{ho} h(t) \right) \tag{4}$$

where $x(t)$ and $y(t)$ are input and output vectors, W_{ih}, W_{hh}, and W_{ho} are weight matrices, and f_h and f_o are hidden and output module activation functions.

LSTM is an artificial neural network method (ANN) used in AI and DL. The LSTM has recurrent connectivity and thus differs from conventional feedforward neural networks (FFNNs). Recurrent neural networks (RNNs) can analyze individ-

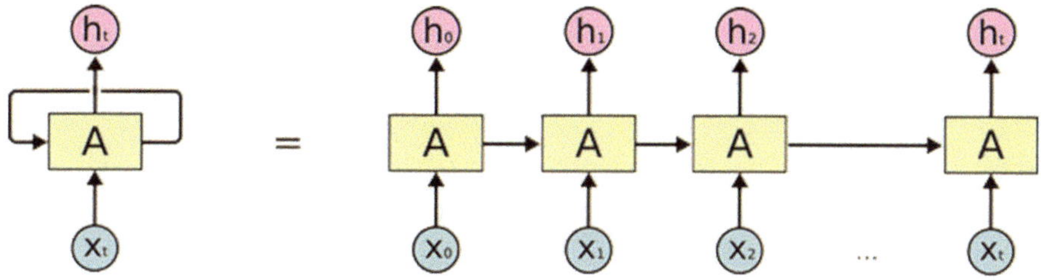

Fig. 2 Schematic of a single-layer recurrent neural network

ual data points and entire data sequences. For example, the LSTM suits tasks such as pattern recognition, machine translation, robot management, video games, health, and wireless networks.

The LSTM is a modified version of a recurrent neural network (RNN) designed specifically for the problem of long-term dependence. They are trained to remember past data and process input data more efficiently in a loop to update the network state over long periods, requiring little to no training. Each recurrent neural network consists of a series of repeating neural network modules. These repeating modules in standard RNNs have a simple structure, such as a single tanh layer. They can more easily remember past data and process input data in a loop to update the network state. The state of the network contains the data from all previous time steps. An LSTM layer learns long-term dependencies between time steps in time series and sequence data. The layer's state comprises the hidden state (also called the output state) and the cell state. The output of the LSTM layer for time step t is stored in the hidden state at time step t.

Using previous time steps as input, an LSTM network can predict future time series or sequence values. Train a regression LSTM network with sequence outputs to predict time series, where the responses (targets) are the training sequences with values shifted by a one-time step. In other words, at each time step of the input sequence, the LSTM network learns to predict the value of the next time step. The vanishing gradient problem in RNN is solved here. LSTM is well suited for classifying, processing, and forecasting time series with unknown duration. It trains the model with backpropagation.

A single LSTM module consists of three gates, an input gate, an output gate, and a forget gate element that facilitates the cell to remember values over time. The gate controls the flow of information to and from the LSTM cell. The hidden state h_t for the LSTM cell can be calculated as follows:

$$i_t = \sigma\left(x_t U^i + h_{t-1} W^i\right) \tag{5}$$

$$f_t = \sigma\left(x_t U^f + h_{t-1} W^f\right) \tag{6}$$

$$o_t = \sigma\left(x_t U^o + h_{t-1} W^o\right) \tag{7}$$

$$\tilde{C}_t = \tanh\left(x_t U^c + h_{t-1} W^c\right) \tag{8}$$

$$C_t = f_t \times C_{t-1} + i_t \times \tilde{C}_t \tag{9}$$

$$h_t = \tanh\left(C_t\right) \times o_t \tag{10}$$

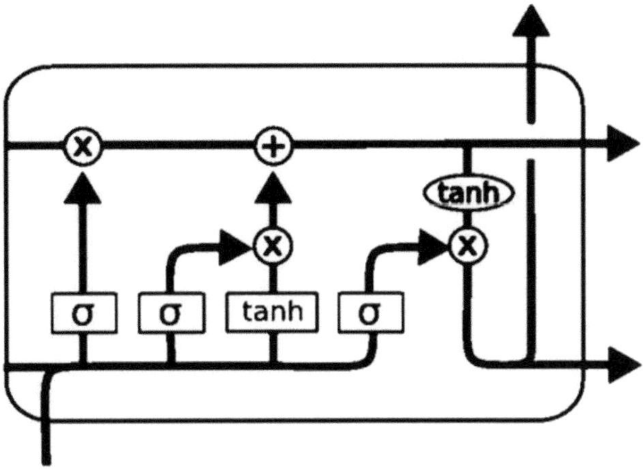

Fig. 3 Structure of the LSTM network

Table 1 Accuracy of delay prediction in IoT

No of epochs	Number of hidden units	Delay prediction	
		MAPE	RMSE
250	100	2.50	0.060
500		1.97	0.051
1000		1.70	0.044
2000		1.20	0.038

Here, *i, f,* and *o* are called the input, forget, and output gates. They have the same equations, only with different parameter matrices (*W* is the current connection between the previously hidden layer and the current hidden layer, and *U* is the weight matrix connecting the inputs to the existing hidden layer). Figure 3 shows the structure of the LSTM network.

4 Simulation Results

This work performed simulations on an Intel (R) Core (TM) I5-3210 M processor running at a CPU 2.50 GHz with 6GB of memory. We created the datasets from the IoT model and used the AnyLogic simulator to simulate the IoT model. After collecting and processing, we divided the dataset into 70% for training and 30% for testing. The MATLAB environment affected the IoT delay prediction with a DL-based LSTM model. The prediction accuracy for IoT delay was measured using the RMSE and MAPE in different cases based on the number of epochs used in training. Table 1 displays the RMSE and MAPE prediction accuracy for IoT traffic.

Table 1 displays the prediction accuracy for M2M network traffic regarding RMSE and MAPE in different cases based on the number of epochs used in training.

Looking at the tabulated results, the model predicted in the case of 2000 epochs has the best prediction accuracy with an RMSE value equal to 0.038 and MAPE

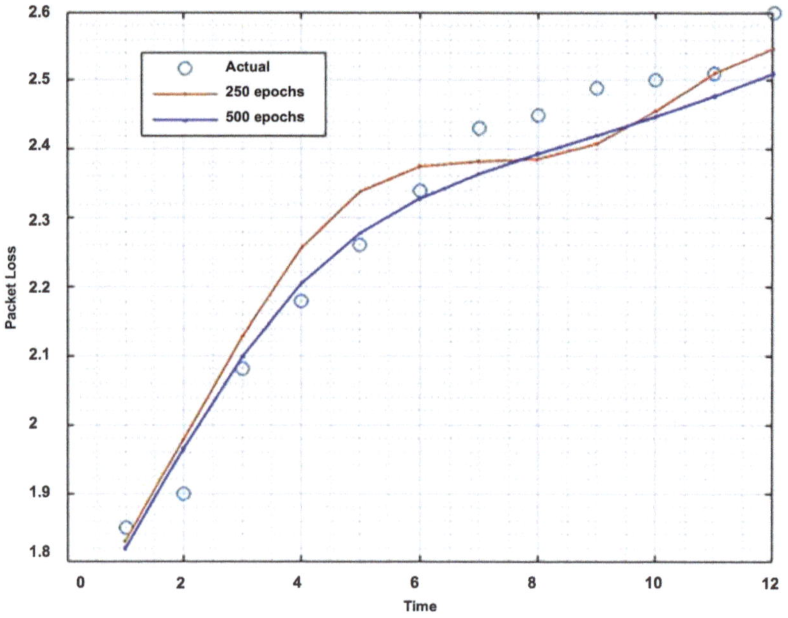

Fig. 4 Prediction models for delays in the M2M network using 250 epochs and 500 epochs

equal to 1.2% compared to the other cases. In this case, the maximum average prediction accuracy improvement is 1.3%. Also, the prediction accuracy in the cases of 500 and 1000 epochs have dropped to RMSE values 0.044 and 0.051, respectively; the MAPE decreased to 1.7% and 1.97%, and the maximum average improvement in prediction accuracy in these cases is 0.8% and 0.53%, respectively.

Figures 4 and 5 show the predicted packet loss using a DL-based LSTM network in different cases, depending on the number of epochs used during the training process. Fig. 4 shows the predicted models when 250 and 500 epochs are used for training. As can be seen in the figure, the predicted models increase with time and deviate slightly from the observed model. As shown in Fig. 5, the 1000 and 2000 epochs are also nearly identical to the observed model.

Figure 6 shows the relationship between prediction error and time. From the figure, the highest prediction error occurs at 5 s when using 250 epochs, while the highest prediction error occurs at 2 s when using 500, 1000, and 2000 epochs.

5 Conclusion

The method for predicting delays in M2M networks using DL and an LSTM network is discussed in this chapter. This method has the advantage of preserving the past state of a time point and providing a more accurate prediction of the next

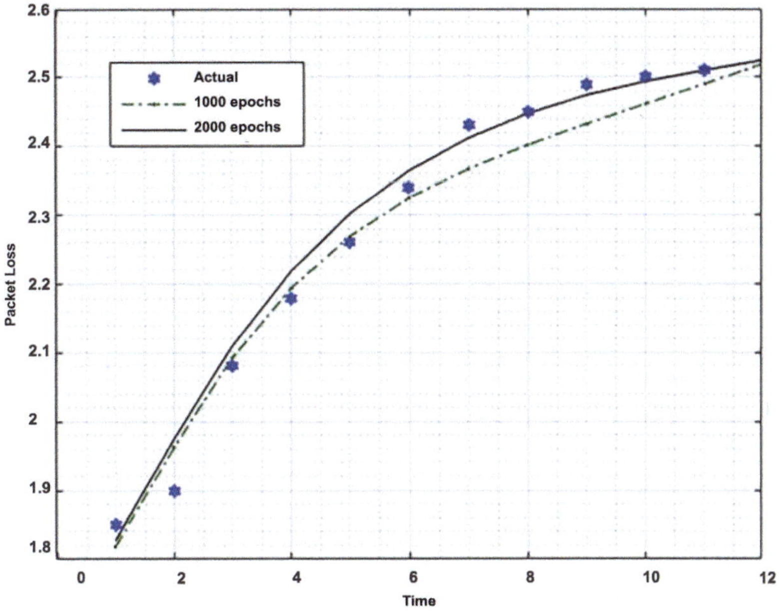

Fig. 5 Prediction models for delays in the M2M network using 1000 epochs and 2000 epochs

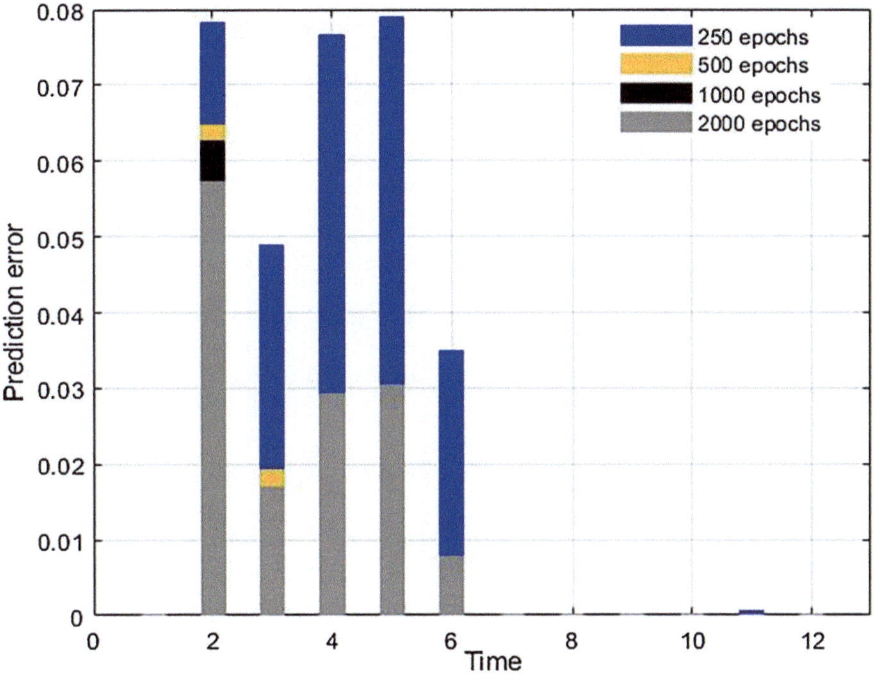

Fig. 6 Prediction error vs time

time point. It also offers benefits over traditional time-series prediction models by optimizing the accuracy of the learning process as training progresses. Including more data makes the model smarter and improves its ability to predict traffic volumes, which is critical for real-time traffic forecasting. RMSE and MAPE are additional accuracy measures used to evaluate the prediction accuracy of the deep learning model. The simulation results show that the model with the epoch number 2000 has the best prediction accuracy compared to the other cases, while the model with the epoch number 500 has poor performance compared to the other cases.

Acknowledgments The research was supported by Applied Scientific Research under the SPbSUT state assignment No. 1022040500653-0, 2023.

References

1. Sarker, I.H.: AI-based modeling: techniques, applications and research issues towards automation, intelligent and smart systems. SN Comput. SCI. **3**, 158 (2022). https://doi.org/10.1007/s42979-022-01043-x
2. Dwivedi, J.N.: Internet of things (IoT) and machine to machine (M2M) communication techniques for cyber crime prediction. In: Pani, S.K., Singh, S.K., Garg, L., Pachori, R.B., Zhang, X. (eds.) Intelligent Data Analytics for Terror Threat Prediction (2021). https://doi.org/10.1002/9781119711629.ch2
3. Abdellah, A., Koucheryavy, A.: Survey on Artificial Intelligence Techniques in 5g Networks. Telecom IT, SPbSUT, Russia, vol. 8, issue. 1, pp. 1–10. http://www.sut.ru/doci/nauka/1AEA/ITT/2020_1/1-10.pdf (April 2020)
4. Abdellah, A.R., Koucheryavy, A.: Artificial Intelligence Driven 5G and Beyond Networks. Telecom IT, SPbSUT, Russia, vol. 10, issue. 2, pp. 1–13. http://www.sut.ru/doci/nauka/1AEA/ITT/2022_2/1-13.pdf (2022)
5. Abdellah, A.R., Mahmood, O.A.K., Paramonov, A., Koucheryavy, A.: IoT traffic prediction using multi-step ahead prediction with neural network. In: IEEE 11th International Congress on Ultra-Modern Telecommunications and Control Systems and Workshops (ICUMT) (2019)
6. Ateeq, M., Ishmanov, F., Afzal, M.K., Naeem, M.: Predicting delay in IoT using deep learning: a multiparametric approach. IEEE Access. **7**, 62022–62032 (2019). https://doi.org/10.1109/ACCESS.2019.2915958
7. Feizimirkhani, R., et al.: Application of Long Short-Term Memory (LSTM) Neural Network for the estimation of communication network delay in smart grid applications. In: 2021 IEEE International Conference on Environment and Electrical Engineering and 2021 IEEE Industrial and Commercial Power Systems Europe (EEEIC/I&CPS Europe), pp. 1–6 (2021) https://doi.org/10.1109/EEEIC/ICPSEurope51590.2021.9584791
8. Abdellah, A.R., Mahmood, O.A., Kirichek, R., Paramonov, A., Koucheryavy, A.: Machine learning algorithm for delay prediction in IoT and tactile internet. Future Internet. **13**(12), 304 (2021)
9. Ateeq, M., Afzal, M.K., Naeem, M., Shafiq, M., Choi, J.-G.: Deep learning-based multi-parametric predictions for IoT. Sustainability. **12**(18), 7752 (2020). https://doi.org/10.3390/su12187752
10. Abdellah, A.R., Alzaghir, A., Koucheryavy, A.: Deep learning approach for predicting energy consumption of drones based on MEC. In: Koucheryavy, Y., et al. (eds.) NEW2AN 2021/ruSMART 2021 LNCS, vol. 13158, pp. 284–296. Springer, Cham (2022). https://doi.org/10.1007/978-3-030-97777-1_24

11. Abdellah, A.R., Volkov, A., Muthanna, A., Gallyamov, D., Koucheryavy, A.: Deep learning for IoT traffic prediction based on edge computing. In: Vishnevskiy, V.M., Samouylov, K.E., Kozyrev, D.V. (eds.) Distributed Computer and Communication Networks: Control, Computation, Communications. DCCN 2020 Communications in Computer and Information Science, vol. 1337, pp. 18–29. Springer, Cham (2020). https://doi.org/10.1007/978-3-030-66242-4_2
12. Abdellah, A.R., Koucheryavy, A.: VANET traffic prediction using LSTM with deep neural network learning. In: Galinina, O., Andreev, S., Balandin, S., Koucheryavy, Y. (eds.) Internet of Things, Smart Spaces, and Next Generation Networks and Systems. (NEW2AN/ruSMART) Lecture Notes in Computer Science, vol. 12525, pp. 281–294. Springer, Cham (2020). https://doi.org/10.1007/978-3-030-65726-0_25
13. Abdellah, A.R., Koucheryavy, A.: Deep learning with long short-term memory for IoT traffic prediction. In: Galinina, O., Andreev, S., Balandin, S., Koucheryavy, Y. (eds.) Internet of Things, SMART Spaces, and Next Generation Networks and Systems (NEW2AN/SMART) Lecture Notes in Computer Science, vol. 12525, pp. 267–280. Springer, Cham (2020). https://doi.org/10.1007/978-3-030-65726-0_24
14. Hameed, A., Violos, J., Santi, N., Leivadeas, A., Mitton, N.: A machine learning regression approach for throughput estimation in an IoT environment. In: 2021 IEEE International Conferences on Internet of Things (iThings) and IEEE Green Computing & Communications (GreenCom) and IEEE Cyber, Physical & Social Computing (CPSCom) and IEEE Smart Data (SmartData) and IEEE Congress on Cybermatics (Cybermatics), pp. 29–36 (2021). https://doi.org/10.1109/iThings-GreenCom-CPSCom-SmartData-Cybermatics53846.2021.00020
15. Ali Shah, S.A., Illanko, K., Fernando, X.: Deep Learning Based Traffic Flow Prediction for Autonomous Vehicular Mobile Networks. In: 2021 IEEE 94th Vehicular Technology Conference (VTC2021-Fall), pp. 01–05 (2021). https://doi.org/10.1109/VTC2021-Fall52928.2021.9625196
16. Khedkar, S.P., Canessane, R.A., Najafi, M.L.: Prediction of traffic generated by IoT devices using statistical learning time series algorithms. Wirel. Commun. Mob. Comput., 5366222., 12 p (2021). https://doi.org/10.1155/2021/5366222
17. Abdellah, A.R., Muthanna, A., Essai, M.H., Koucheryavy, A.: Deep learning for predicting traffic in V2X networks. Appl. Sci. **12**(19), 10030 (2022). https://doi.org/10.3390/app121910030
18. Celestine, I., Kashif, B.A., Atharva, P., Sujatha, R., Moy, C.J., Swetha, P., Rishita, M., Sofia, P., Ohyun, J.: COVID-19 patient health prediction using boosted random Forest algorithm. J. Front. Public Health. **8** (2020) https://www.frontiersin.org/articles/10.3389/fpubh.2020.00357
19. Khedkar, S.P., Canessane, R.A., Najafi, M.L.: Prediction of traffic generated by IoT devices using statistical learning time series algorithms. Hindawi Wirel. Commun. Mob. Comput. **2021**, 1–12 (2021). https://doi.org/10.1155/2021/5366222
20. Abdellah, A.R., Abdulkareem Mahmood, O., Koucheryavy, A.: Delay prediction in IoT using machine learning approach. In: 2020 12th International Congress on Ultra Modern Telecommunications and Control Systems and Workshops (ICUMT), pp. 275–279 (2020). https://doi.org/10.1109/ICUMT51630.2020.9222245
21. Abdellah, A.R., Mahmood, O.A.K., Paramonov, A., Koucheryavy, A.: IoT traffic prediction using multi-step ahead prediction with neural network. In: 2019 11th International Congress on Ultra Modern Telecommunications and Control Systems and Workshops (ICUMT), pp. 1–4 (2019). https://doi.org/10.1109/ICUMT48472.2019.8970675
22. Volkov, A., Abdellah, A.R., Muthanna, A., Makolkina, M., Paramonov, A., Koucheryavy, A.: IoT traffic prediction with neural networks learning based on SDN infrastructure. In: Vishnevskiy, V.M., Samouylov, K.E., Kozyrev, D.V. (eds.) Distributed Computer and Communication Networks. DCCN 2020 Lecture Notes in Computer Science, vol. 12563, pp. 64–76. Springer, Cham (2020). https://doi.org/10.1007/978-3-030-66471-8_6

Energy-Efficient Beam Shaping in MIMO System Using Machine Learning

Aigul R. Absalyamova, Anna V. Voronkova, Ekaterina A. Lopukhova, Elizaveta P. Grakhova, Rosalina V. Ishakova, and Grigoriy S. Voronkov

1 Introduction

Currently, against the backdrop of a rapidly developing infrastructure of telecommunications networks, network traffic and the number of connected subscribers are growing. Also, user demands for performance continue to grow, which leads to an increase in network load. The Cisco Annual Internet Report showed that by 2023, almost two-thirds of the world's population will have Internet access, with a total of 5.3 billion people (66% of the world's population), compared to 3.9 billion (51% of the world's population) in 2018. While the entire number of the all global mobile subscribers will grow up from 66% to 71% of the total population of the Earth [1].

This work was funded under the grant of the Russian Science Foundation (Project № 075-03-2021-014).

A. R. Absalyamova
Department of Telecommunication Systems, Ufa State Aviation Technical University, Ufa, Russian Federation

A. V. Voronkova · E. A. Lopukhova · E. P. Grakhova · G. S. Voronkov (✉)
School of Photonics Engineering and Research Advances (SPhERA), Ufa State Aviation Technical University, Ufa, Russian Federation
e-mail: voronkova.av@ugatu.su; lopukhova.ea@ugatu.su; grakhova.ep@ugatu.su; voronkov.gs@ugatu.su

R. V. Ishakova
Department of High Performance Computing Technologies and Systems, Ufa State Aviation Technical University, Ufa, Russian Federation
e-mail: ishakova.rv@ugatu.su

A. A. Abd El-Latif et al. (eds.), *Secure Edge and Fog Computing Enabled AI for IoT and Smart Cities*, EAI/Springer Innovations in Communication and Computing,
https://doi.org/10.1007/978-3-031-51097-7_8

With an increase in the volume of transmitted information, an increase in energy costs is required to provide the necessary signal-to-noise ratio for each subscriber. Consequently, with an increase in the demand for communication services, the need increases for antenna systems, which will provide the necessary energy budget with minimization of energy costs without loss of quality of the services provided. The problem of energy consumption of the base station (BS) of the latest generation is central to many scientific studies. For example, it is shown that the standard power consumption of 5G BS is about 3500 W [2], while a 4G base station consumes two to three times less. Therefore, the issue of reducing the energy consumption of the latest generation networks is relevant in our time.

In 5G networks, MU-MIMO (multiuser multiple input multiple output) technology is widely used to provide an individual level of service to subscribers [3]. One of the ways to efficiently use the energy resources of the MU-MIMO system is the formation of radiation patterns (RP) in such a way that not all elements of the antenna array are involved, but only those necessary for a given number of subscribers. To optimize this process, it is advisable to consider the option of introducing machine learning (ML) and neural networks (NN) into antenna systems.

A neural network with a supervised learning algorithm is discussed [4], which presents a hybrid beamforming algorithm that seeks to maximize the achievable total data rate in a MIMO system in the millimeter wave. The same ML algorithm is used [5] and [6], where hybrid analog-to-digital beamforming is used for millimeter-wave communication.

The problem of efficient use of energy resources using ML is solved in the paper [7], where an ML algorithm for the antenna system is developed. This algorithm generates clusters using the K-means clustering algorithm and an algorithm based on a mixed Gaussian model. When the algorithm is applied, users in each cluster are served by only one BS, closest to the center of the cluster. This significantly reduces interference from other devices and reduces the computational complexity of the central device.

Application of these algorithms in antenna systems can provide better performance. For example, when the maximum transmit power is 20 dBm, the energy efficiency of traditional communication technologies reaches its maximum, and the energy efficiency when using the K-means clustering algorithm and the Gaussian mixed model clustering algorithm is about 230% and 180% higher than the energy efficiency of the traditional communication model in the antenna system.

Due to the relevance and potential of methods for reducing the power consumption of communication systems through intelligent beamforming [8], it is reasonable to use machine learning to solve the above issue, namely the K-means clustering algorithm, since it shows high energy efficiency and does not require high computational base station costs. The disadvantages of this algorithm include the fact that it requires some a priori information. However, in this work, this is rather an advantage, since, by changing the initial data, it is possible to study various dependencies, for example, the dependence of the number of subscribers on the number of elements of a phased antenna array (PAR).

2 Beam-Shaping App

The result of this work is an application developed in the Python environment. The block diagram is shown in Fig. 1.

The a priori data for the operation of the algorithm are the number of elements of the antenna array N_{el}; this indicator is necessary since it is necessary to establish the boundaries of the number of elements of the phased array. The distribution of subscribers for different levels of service is created by generating their coordinates with a certain sparsity (dispersion) relative to demand points.

Service level was determined in this case as required bandwidth for serving subscribers' traffic. By default, unless other ratios are specified, it is assumed in this work that for the first level, the number of demand points is 35% of the total number of demand points, for the second 35%, for the third 20%, and for the fourth 10%.

Calculate the distance between the subscriber and the BS using the formula:

$$d = \frac{0.8}{2000}\sqrt{(x_{BS} - x_{\mathrm{AB}})^2 + (y_{BS} - y_{\mathrm{AB}})^2}, \tag{1}$$

where x_{BS} and y_{BS} are the coordinates of the BS along the x-axis and the y-axis, respectively, x_{AB} and y_{AB} are the subscriber's coordinates along the x-axis and the y-axis, respectively. Determining the distance is necessary to calculate the attenuation of the signal along the propagation path. In this work, the Longley-Rice model [9, 10] is used to estimate it. The location of subscribers is determined using cell-id technology.

For multilevel clustering of subscribers by service level and distance, the abovementioned K-means algorithm is used. The input data are the coordinates of the subscribers and the levels of service they need.

Efficiency was assessed using the adjusted Rand index – Adjust_rand_s (tARI) (a measure of similarity with the real situation) and the Homogeneity_s (thomo) efficiency indicator (Table 1).

As it can be seen from the Table 1, the values of the tARI and thomo indicators decrease with an increase in the number of subscribers; however, they remain quite large and close to one, which makes it possible to judge that the clustering algorithm is quite consistent with real conditions.

Visualization of clustering for 800 subscribers and 80 demand points is shown in Fig. 2.

The result of the algorithm's work at the current stage is the radiation patterns constructed for each service level. The application generates about several dozen radiation patterns for all levels in total.

Next, we calculated the ratio C for the channel bandwidth B as the ratio of the electrical signal level to the noise level (SINR). In this case, the throughput is multiplied by the number of subscribers N_{AB}, since the calculation is made for a cluster, and not for a single subscriber:

$$\mathrm{SINR} = 2^{C \bullet N_{\mathrm{AB}}/B}. \tag{2}$$

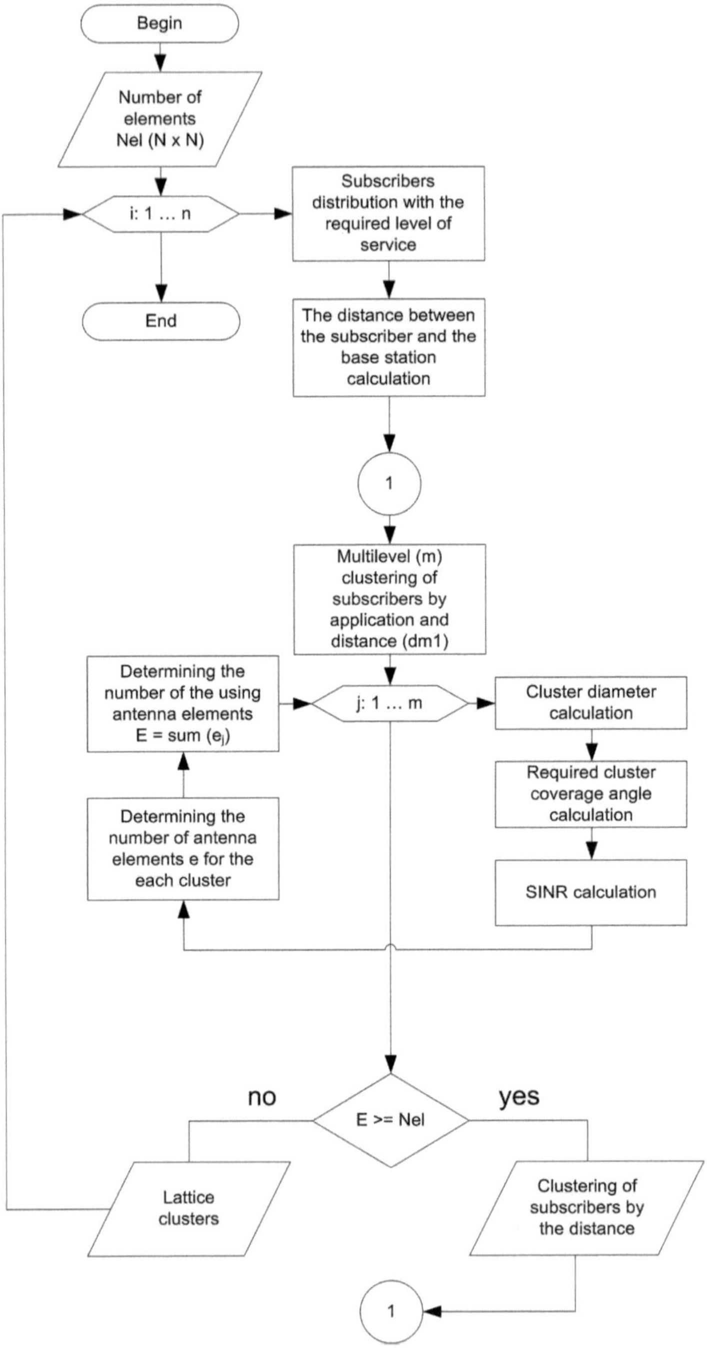

Fig. 1 Block diagram of the application algorithm

Table 1 Performance indicators of the clustering algorithm

Number of subscribers	tARI	thomo
250	0.917327	0.950919
500	0.892469	0.945836
1000	0.889289	0.926496
1350	0.880368	0.921034

The bandwidth itself takes values of 2 Mbps, 5 Mbps, 12 Mbps, and 35 Mbps according to each level of service. The channel bandwidth is assumed to be average for 5G networks and equal to 100 MHz. The cycle will continue until the desired SINR is obtained for each cluster.

Further, the summation of all involved PAR elements for each service level occurs.

After the completion of the clustering cycle, the condition "$E >= N_{el}$" is checked, where the number of PAR elements is compared with the given limits on the number of elements. If the received number of elements exceeds the limits, then the clustering cycle is repeated anew, but after changing the initial data.

The result of the algorithm is synthesized radiation patterns, number of involved elements of the phased array, and visualization of the operation of the clustering cycle.

The entire cycle of the algorithm is repeated for each of the service levels.

3 Conclusion

Thus, an application was developed that allows one to calculate the required number of elements of a phased antenna array to increase the energy efficiency of the system. In addition to calculating the required number of elements, the application calculates energy efficiency with an increase in the directivity factor and the output power of the BS per sector.

A method for increasing energy efficiency through intelligent control of the antenna pattern is proposed. Management is based on the clustering of subscribers by service levels and coordinates. K-means was chosen as the clustering algorithm, as it has the following advantages: speed and ease of implementation. The proposed solution makes it possible to increase the energy efficiency of the 5G BS by an average of 38%, which can save up to 1449 kWh of energy.

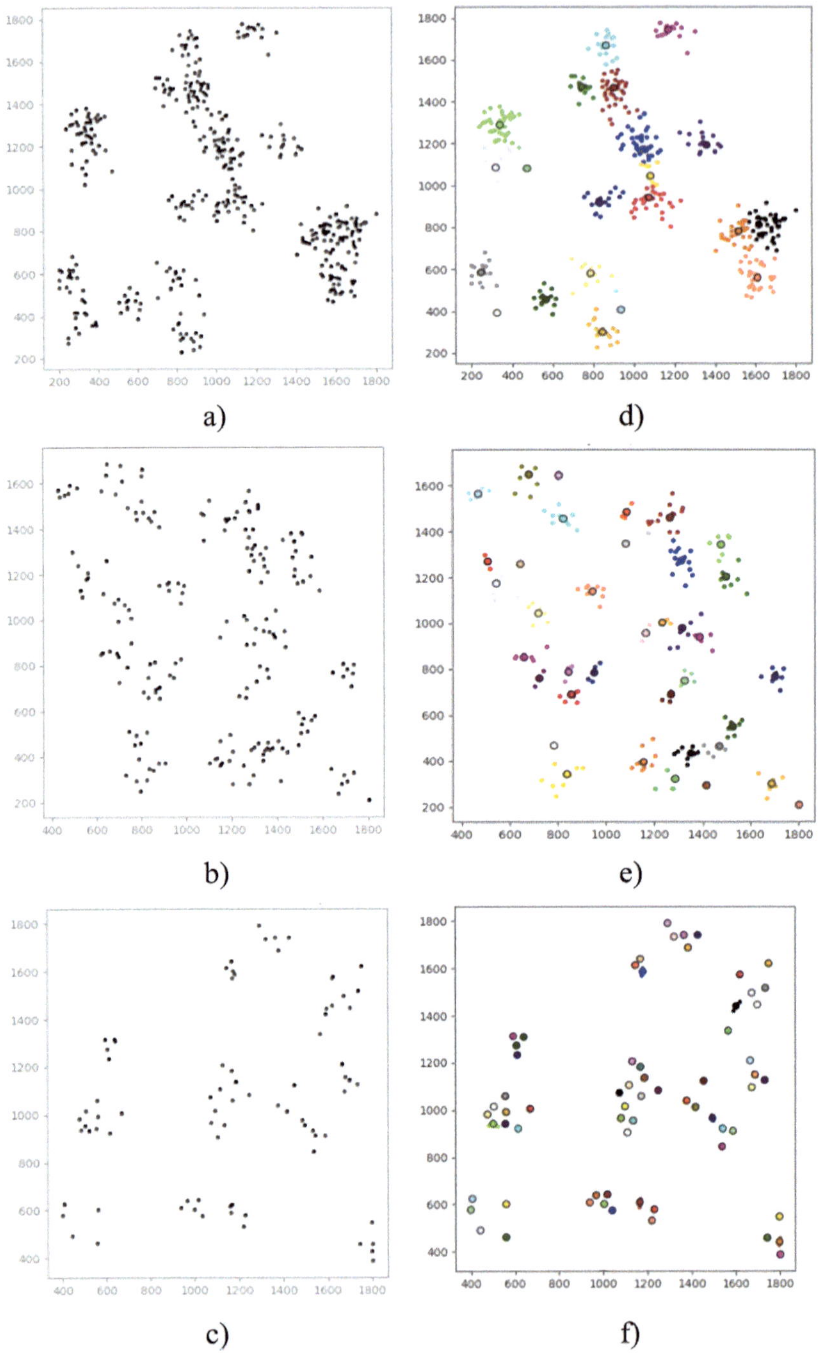

Fig. 2 Visualization of subscriber clustering for service levels: (**a**) non-clustering location for first service level; (**b**) non-clustering location for second service level; (**c**) non-clustering location for third service level; (**d**) clusterization for first service level; (**e**) clusterization for second service level; (**f**) clusterization for third service level

Acknowledgments The study was carried out as part of the work on the state assignment of the Ministry of Education and Science of Russian Federation for the FSBEU of HE "USATU" (scientific theme code #FEUE-2021-0013, agreement № 075-03-2021-014) in the youth research laboratory of the SEC Sensor systems based on integrated photonics devices».

References

1. https://www.cisco.com/c/en/us/solutions/collateral/executive-perspectives/annual-internet-report/white-paper-c11-741490.html [Electronic resource]
2. Wang, Y.: Energy-saving scheme of 5G base station based on LSTM neural network. In: 2nd International Conference on Applied Physics and Computing (2021)
3. Sampson, N., Erfanian, J., Hu, N.: NGMN 5G White Paper 2, 32. Retrieved 2020
4. Lizarraga, E., Maggio, G., Dowhuszko, A.: Hybrid beamforming algorithm using reinforcement learning for millimeter wave wireless systems. In: 2019 XVIII Workshop on Information Processing and Control (RPIC). IEEE, pp. 253–258 (2019)
5. Yetis, C., Bjornson, E., Giselsson, P.: Joint analog beam selection and digital beamforming in millimeter wave cell-free massive MIMO systems. IEEE Open J. Commun. Soc. **2**, 1647–1662 (2021)
6. ElHalawany, B.: Leveraging machine learning for millimeter wave beamforming in beyond 5G networks. IEEE Syst. J. (2021)
7. He, C. et al.: Energy Efficient Power Allocation Based on Machine Learning Generated Clusters for Distributed Antenna Systems (2019)
8. Lovato, R., Gong, X.: Phased antenna array beamforming using convolutional neural networks. In: IEEE International Symposium on Antennas and Propagation and USNC-URSI Radio Science Meeting, APSURSI 2019 – Proceedings. Institute of Electrical and Electronics Engineers, pp. 1247–1248 (2019)
9. Oughton, E.: Longley-Rice Irregular Terrain Model (itmlogic) (Version 1). [Source code]. . https://github.com/edwardoughton/itmlogic (2020)
10. Oughton, E., Russell, T., Johnson, J., Yardim, C., Kusuma, J.: The irregular terrain model by Longley and rice. J. Open Source Softw. (2020)

Channel Cluster Configuration Selection Method for IEEE 802.11 Network Planning

Anton Vikulov and Alexander Paramonov

1 Introduction

During the design process of IEEE 802.11 networks [1] including IEEE 802.11ax [2], which are deployed according to the cellular principle, it is often necessary to correctly select a specific channel plan. This may be required from regulatory specifics and to justify the design for the customer in order to provide required throughput values for specific client devices. The planning of wireless networks of various standards was considered in the literature [3]. The main sources of interference for the IEEE 802.11 network are adjacent networks of the same standard, as well as access points (APs) of the same network operating at adjacent channels in the spectrum, which creates a negative impact on the AP.

Several works [6, 7] were devoted to the study of planning [4] of wireless local area networks (WLAN) of the IEEE 802.11 standard for various applications [5]. In these works, a probabilistic approach has been considered in order to provide a placement of APs on a plane and calculate the total interference spectrum. In this chapter, this problem will be considered from a different perspective.

Previously a model [8] was proposed in order to provide a geometrical approach, which consists of constructing a tessellation structure on the plane with AP coverage zones, including the choice of a lattice (i.e., the type of translational symmetry defined by the basis) and cell unit. In this case, we will use crystallographic terms to explain the geometric essence of the proposed approach. So, under the cell (or elementary) unit [9] of the structure, we understand a geometrical figure, which is presented by a group of AP coverage areas associated with each of the nodes of the

A. Vikulov (✉) · A. Paramonov
The Bonch-Bruevich Saint-Petersburg State University of Telecommunications, Saint-Petersburg, Russia
e-mail: asv012016@gmail.com

A. A. Abd El-Latif et al. (eds.), *Secure Edge and Fog Computing Enabled AI for IoT and Smart Cities*, EAI/Springer Innovations in Communication and Computing,
https://doi.org/10.1007/978-3-031-51097-7_9

lattice. In the spectral sense, the cell unit corresponds to the concept of a channel cluster [10] while, however, focusing on the geometry of the tessellation structure. In other words, by translating (mapping) the cell unit to each of the lattice nodes, we obtain a tessellation structure. A model of adjacent channel interference from the AP of its own network was proposed, based on the designated geometrical approach [11].

The purpose of this work is to develop a method that will allow to choose the best channel configuration for specific cell unit, i.e., provide a correspondence between the central frequency of a channel of a given type and the AP number in a cell unit in such a manner that the characteristics of the radio coverage of the entire resulting structure would be the best. Let us now consider this issue in a more detailed view.

2 Problem Statement

For the quality criterion of radio coverage, which determines the choice of modulation and coding sequence (MCS), we will consider the signal-to-noise ratio (SNR) measured on the side of the access point in the chosen cell unit. For example, Fig. 1 shows a plane tessellation example for a channel cluster of size M equal to 4, in a structure with a coordination number N (the number of elements at a minimum distance from a given one) in the plane equal to 4. Translational symmetry vectors (basis) $\boldsymbol{a}, \boldsymbol{b}$ are marked with red, and the target cell unit is located at the node O of the lattice with coordinates (0;0).

Let D denote the size of the grid taken into account in the calculations. Figure 1 shows the structure of 25 cell units in the tessellation formed when the grid size D is equal to 2. Here, 24 cell units are the results of translations of the original unit along both axes in the range:

$$D \in [-2; 2]$$

We will further consider space tessellation structures and estimate parameters of the channel cluster which has a three-dimensional structure. This is important for WLAN design in many possible cases, for example, for a multistoried building.

The total number of possible channel configurations in a cluster is $M!$ Now we may formulate the problem in the following way. For the infinite tessellation structure, it is necessary to find the best channel configuration, i.e., a one-to-one correspondence between the AP number of the cell unit and the center frequency of the IEEE 802.11 channel of a given type, which would meet the requirements of coverage characteristics.

We will consider a model of adjacent channel interference proposed [11]. Let us list the requirements for channel configuration of a cell unit from the standpoint of the SNR value for every AP in it:

1. Average SNR for all APs should be maximum.

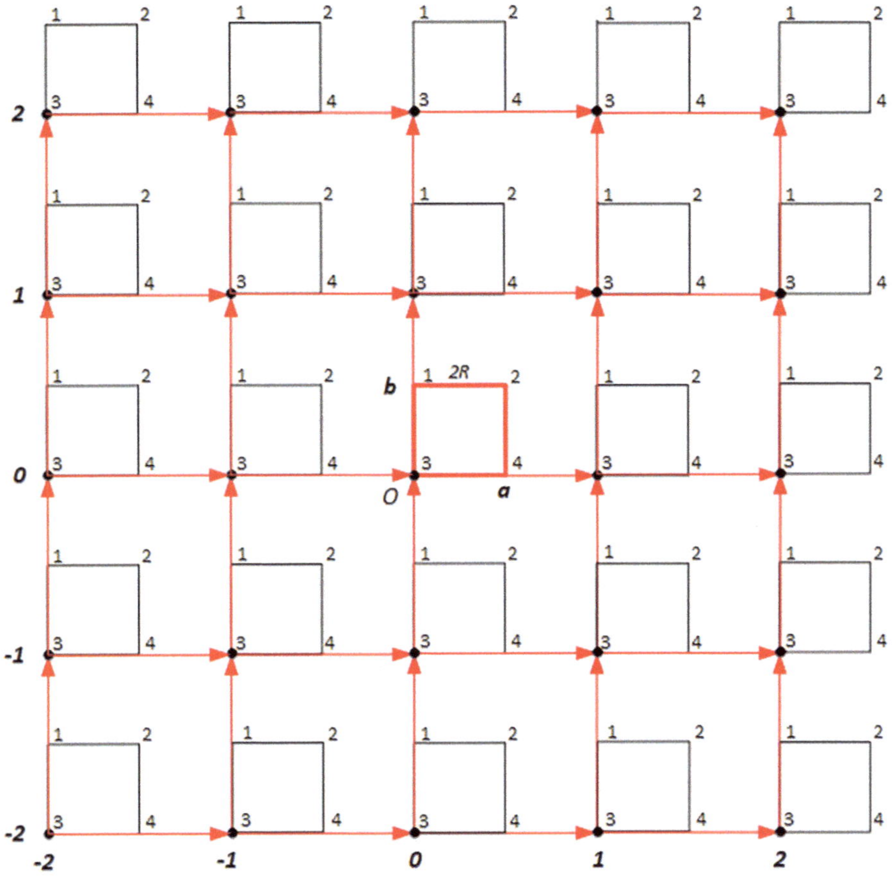

Fig. 1 Tessellation example for M = 4, N = 4, and D = 2

2. The max(min) condition for the SNR must be satisfied.
3. The third largest SNR should be the maximum. This condition is relevant in the case of a uniform distribution of channels in a continuous section of the spectrum.
4. The minimum SNR must not be lower than the allowable one. This is the most typical design requirement.

These requirements will be discussed in detail later.

Let us make the following assumptions:

1. The coverage area is infinite. It makes it possible to compare different solutions to the problem of finding the best channel configuration.
2. The coverage area formed by every access point is a circle of radius R.
3. The distance between any two nearest access points is $2R$.
4. The frequency plan uses channels of the same type, for example, HE20.

5. Q is the probability of transmission in the channel and is the same and equal for all channels.
6. When constructing the total spectrum of adjacent channel interference, we will take into account the influence of the total spectrum of interference from other access points on the side of the considered AP.

Our goal is to construct a method for choosing the best frequency configuration out of all theoretically possible ones.

3 Channel Configuration Selection Method

If the size of the frequency cluster is equal to M, then there will be M access points in the cell unit of the tessellation structure, each operating on its own channel. Then the number of possible channel configurations will be $M!$, i.e., the number of possible configurations is determined by possible permutations out of M. Let us put each AP in correspondence with some central frequency of the channel of a predetermined type. We will define a matrix-vector F containing the central frequencies of the channels specified by the conditions of the problem.

$$F = (F_1\ F_2 \cdots F_M) \quad \text{(MHz)} \tag{1}$$

The matrix H containing all possible frequency configurations includes all possible permutations of the vector F:

$$H = \begin{pmatrix} F_{1,1} & F_{1,2} & \cdots & F_{1,M!} \\ F_{2,1} & F_{2,2} & \cdots & F_{2,M!} \\ \cdots & \cdots & \cdots & \cdots \\ F_{M,1} & F_{M,2} & \cdots & F_{M,M!} \end{pmatrix} \quad \text{(MHz)} \tag{2}$$

Now each possible channel configuration out of their total number $M!$ for the cell unit will be uniquely specified by the column of the matrix H.

The proposed configuration selection method includes five steps.

Step 1. Formation of initial data.

In this step, specific quantitative values of the initial variables and the type of the tessellation structure are determined. The initial data in this problem includes:

- Radius R of the coverage area formed by the AP.
- Signal attenuation model $L(f,d)$ that defines the signal attenuation L (dB) as a function of frequency f (MHz) and distance to source d.
- Lattice type and its basis.
- Geometry of the cell unit of structure.
- N – coordination number of the structure.

- M – size of the channel cluster, equal to the number of APs in the cell unit.
- F – matrix-vector containing center frequencies of the channels used in the channel plan.
- Type of channels used in the channel plan.
- H – matrix of possible channel configurations.
- Q – probability of channel occupation.
- NF – background noise, dBm.
- P_T – required signal strength level, dBm.
- P_{rad} – transmission power of all APs radios in the considered range.

Step 2. Calculation of the matrix of possible solutions.

The matrix $SNR_{n,m}$, (n is the vertex number and m is the solution number), containing a set of SNR values, is calculated according to the model proposed [11]. The calculation is performed for all possible channel configurations of a cell unit of a given structure and size for several different grid sizes D.

At the same time, the SNR values themselves are calculated according to the formula:

$$SNR_{n,m} = 10\log_{10}\frac{\int_{F_T-A}^{F_T+A} S_{Tw}(f)df}{\int_{F_T-A}^{F_T+A} W_{Iw}(f)df}\ (\text{dB}) \tag{3}$$

where

F_T is the center frequency of the signal, MHz;
A is the half-width of the signal spectral mask section, MHz.

In this case, the size of the matrix is determined by the following ranges of index values.

In formula (3), the total spectrum of all adjacent channel interference W_{Iw} from cells of the same WLAN is calculated according to:

$$W_{Iw}(f) = NF + Q\sum\nolimits_{i=-D}^{D}\sum\nolimits_{j=-D}^{D}\sum\nolimits_{k=-D}^{D}\sum_{r=1}^{M} US_{Iw}\left(H_{k,m}, d_{i,j}(n), f\right) \tag{4}$$

while:

$$U = \begin{cases} 1, & r \neq n \\ 0, & r = n \end{cases},$$

where

U is the function indicator of adjacent channel interference;
i,j,k are the coordinates in the system of translational symmetry (Fig. 1);

D is the grid size;
S_{Iw} is the spectral mask of a channel of a given type;
$H_{n,m}$ is a matrix of possible frequency configurations for a given cell unit, which contains n APs.

The spectral signal mask is set according to:

$$S_{Tw}(f) = 10^{\frac{P_T + S(f - F_T)}{10}} \text{ (mW)} \tag{5}$$

Spectral mask of the interfering signal is given by:

$$S_{Iw}(f) = 10^{\frac{P_{rad} + G_t + S(f - F_I) - L(d)}{10}} \text{ (mW)}, \tag{6}$$

where

F_I is the center frequency of the interfering signal channel, MHz;
G_t is the gain of the AP transmitting antenna, dB;
L(d) is the interfering signal attenuation depending on the distance d to the source, dB.

Note that the choice of a specific attenuation model [12–16] does not have an impact on the choice of the best channel configuration but only affects the quantitative characteristics.

Step 3. Construction of a variational series of solutions.
In order to select the best solution, the variational series is constructed with double sorting in ascending order – firstly by the average signal-to-noise ratio and secondly by the minimum one, and thus it contains two values for each solution.
Step 4. Finding the optimal solution (or group of solutions).

At this step, having $M!$ sets of channel configurations, it is necessary to find out such a solution m (or a group of solutions) where the optimality conditions would be satisfied.

Taking into account the previously named requirements, we formulate the problem of finding the best channel configuration in the following way. Within the cell unit, such a channel configuration of the APs must meet the following conditions:

1.

$$\max \left\langle SNR_{n,m} \right\rangle, \text{dB}$$

2.

$$\max\ \min \left\langle SNR_{n,m} \right\rangle, \text{dB}$$

3.

$$\max\langle SNR_{3,m}\rangle, \text{dB}$$

4.

$$\min\langle SNR_{n,m}\rangle > 25 \text{ dB (for example)}$$

The first condition requires maximizing the average value of SNR_n among all possible m. This is necessary to provide the best average SNR for all APs in the cell unit and, as a result, better transmission rates at higher MCS.

The second condition requires the absence of "failed" APs in the cell unit in terms of the SNR value, i.e., the minimum SNR in the cell unit must be the maximum among all possible solutions. This is necessary to exclude from the list of the best solutions cases in which a relatively high average SNR is theoretically possible with the presence of a low value (possibly on only one AP) minimum SNR.

The third condition is optional and is related to the peculiarities of channel placement in the spectrum. In any of the problems under consideration, two channels located at the edges of the spectrum will have advantages over the others due to the fact that for them the interference will be "one sided" in the spectral sense. Therefore, the channel with the third largest SNR value will be one of the determining ones in each channel configuration. Let us denote the SNR of such an AP as $SNR_{3,m}$. In some sense, it can be considered to be the "best" channel configuration, since the quality of its work will be determined to a greater extent by the geometry of the cell unit and channel configuration, and not by the fact of being located on the edge of the considered part of the spectrum.

The fourth condition is optional and can define a lower bound on the SNR of the AP of the cell unit since this condition usually appears explicitly when solving specific design problems. In this case, the value is given as an example, and the specific value may depend on the conditions of the design problem. In reality, this condition is the most typical for any WLAN design requirement.

Step 5. Extrapolation of the solution to (grid extrapolation).

Since the average SNR is calculated for a number of possible values of the grid size D, and the calculation for large D is difficult due to the high computational complexity, it seems convenient to apply the following approach. The value at large D will have a limit associated with the fact that the number of cell units in the solution, whose influence on the target AP is considered, becomes so large for the considered solution geometry that it becomes indistinguishable from infinity. At the same time, a further increase in D does not have a significant effect on the result due to the large distance to the interference. In such a case, the average SNR from step 1 for the infinite space case is:

$$\langle SNR_{n,m}\rangle_{\infty} = \lim_{D\to\infty} \left(\max\langle SNR_{n,m}\rangle\right) \text{ (dB)} \tag{7}$$

To find such a limit based on the calculated SNR values for the found solutions, it is convenient to use the following approach. We propose the following approximating function:

$$\max\langle SNR_{n,m}\rangle = u_1 \bullet e^{U_2-D} + \langle SNR_{n,m}\rangle_{\infty} \text{ (dB)} \tag{8}$$

where u_1 and u_2 are some coefficients selected by the least squares method.

It can be seen from the formula that function (8) asymptotically approaches the value for large *D*. This is important since in further work it will be shown that not all cases relevant to practice allow an acceptable computation time for sufficiently large *D*.

Similarly, we extrapolate the minimum value of $\min(SNR_n)$ for the best solution.

Note that the solutions for large *D*, corresponding to the large size of the tessellation structure, are a geometric simplification, since in practice at such distances the earth's surface, the occupancy of the Fresnel zone, and, as a result, the height of the AP installation start to give its impact. So, for example, in the flat case for $D = 100$, with $M = 3$ and $N = 3$, with $R = 10$ m, the linear size of the total tessellated space will be about 3.5 km, which is several times higher than the practical operating range of the IEEE 802.11 AP. However, it is at large *D* that the differences in the characteristics of the selected frequency configurations will be most clearly visible.

In future work, we will consider specific solutions using the example of cases with several *M* encountered in practice.

4 Conclusions

1. Based on the model of adjacent channel interference in regular structures, a method is proposed for the channel planning of the IEEE 802.11 networks, which makes it possible to select the best channel configuration in terms of SNR and thus to accurately define the capabilities of the selected frequency plan.
2. Calculations of all possible solutions for a high number of channels are highly resource demanding. Thus a simplifying approach should be proposed.
3. The grid extrapolation concept has been proposed in order to precisely estimate the difference between different possible channel configurations.
4. In cases where it is possible to determine the best frequency configuration, the solution (solution group) does not depend on the signal attenuation model.

In the development of this work, a calculation will be performed for frequency plans with several possible channels, as well as an analysis of the resulting solutions.

References

1. Institute of Electrical and Electronics Engineers. 802.11-2020. IEEE Standard for Information Technology. Telecommunications and Information Exchange between Systems. Local and Metropolitan Area Networks. Specific Requirements. Part 11. Wireless LAN Medium Access Control (MAC) and Physical Layer (PHY) Specifications. IEEE (2021). https://doi.org/10.1109/IEEESTD.2021.9363693
2. Institute of Electrical and Electronics Engineers. 802.11ax-2021. IEEE Standard for Information Technology. Telecommunications and Information Exchange between Systems Local and Metropolitan Area Networks. Specific Requirements. Part 11. Wireless LAN Medium Access Control (MAC) and Physical Layer (PHY) Specifications Amendment 1: Enhancements for High-Efficiency WLAN. IEEE (2021). https://doi.org/10.1109/IEEESTD.2021.9442429
3. Westcott, D.A.: CWAP Certified Wireless Professional Official Study Guide (PW-270) (D.A. Westcott, D.D. Coleman, P. Mackenzie, B. Miller), 696 p. Wiley (2011)
4. Goldsmith, A.: Wireless Communications, 427 p. A. Goldsmith. Cambridge University Press (2005)
5. Paramonov, A.: Clustering optimization for out-of-band D2D communications. In: Paramonov, A., Hussain, O., Samouylov, K., Koucheryavy, A., Kirichek, R., Koucheryavy, Y. (eds.) Wireless Communications and Mobile Computing, vol. 2017. https://doi.org/10.1155/2017/6747052
6. Makolkina, M., Vikulov, A., Paramonov, A.: The augmented reality service provision in D2D network. Commun. Comput. Inf. Sci. **700**, 281–290 (2017)
7. Tonkikh, E.V., Paramonov, A.I., Koucheryavy, A.E.: Modeling the network structure of the IoT using fractals. Electrosvyaz. **4**, 55–62 (2021). https://doi.org/10.34832/ELSV.2021.17.4.007
8. Vikulov, A.S., Paramonov, A.I.: Arrangement of standard structures for tiling the plane for frequency and area planning of IEEE 802.11 networks. Radio Telecommun. Syst. **2**(41), 17–28 (2021)
9. Nesse, D.W.: Introduction to Mineralogy. Oxford University Press, New York (2000) 442 p
10. Nurelmadina, N., Nafea, I., Younas, M.: Evaluation of a channel assignment scheme in mobile network systems. Hum. Cent. Comput. Inf. Sci. **6**, 21 (2016). https://doi.org/10.1186/s13673-016-0075-0
11. Vikulov, A., Paramonov, A., Tatarnikova, T.: Geometrical approach to the plane tessellation in the IEEE 802.11 networks channel planning. In: Internet of Things, Smart Spaces, and Next Generation Networks and Systems. NEW2AN ruSMART 2021 Lecture Notes in Computer Science, vol. 13158, pp. 449–469. Springer, Cham (2021). https://doi.org/10.1007/978-3-030-97777-1_38
12. Rec. ITU-R P.525-2 Calculation of free-space attenuation (1994)
13. Rec. ITU-R P.1238-8 Propagation data and prediction methods for the planning of indoor radiocommunication systems and radio local area networks in the frequency range 300 MHz to 100 GHz (2016)
14. Rec. ITU-R P.676-6 Attenuation by atmospheric gases (2005)
15. Rec. ITU-R P.530-12 Propagation data and prediction methods required for the design of terrestrial line-of-sight systems (2007)
16. Rec. ITU-R P.833-9 Attenuation in vegetation (2016)

Service Migration Algorithm for UAV Recharge Zones in Future 6G Network

Vadim Kovolenko, Abdelhamied A. Ateya, Ammar Muthanna, and Andrey Koucheryavy

1 Introduction

In 1991, Kevin Ashton, in his presentation for Procter & Gamble on the impact of Radio-Frequency Identification (RFID) tags on various market areas, may have first used the term "Internet of Things" [1]. Ericsson (in 2009) and Cisco (in 2010) companies carried out their own analyses on the "Internet of Things" market development. As a result of research conducted by these companies, it was found that the number of Internet of Things devices will reach 50 billion by 2020 [2, 3]. Of course, their forecasts did not come true, because, for example, by 2018, according to the "Internet of Things" market analysis, performed by Strategy Analytics, the number of devices connected to the Internet reached only 22 billion [4]. However, the development of the "Internet of Things" concept until now has high prospects. So, for example, according to new forecasts made by the same company Strategy Analytics, the number of Internet of Things devices will reach 50 billion by 2030. In addition, the development of the "Internet of Things" concept contributed to the emergence of ultrahigh density networks and networks of the fifth-generation 5G/IMT-2020. As the number of devices connected to the network increases, so does the total amount of traffic they receive. At a certain point in the development of the existing basic infocommunication infrastructure, it will not be enough to fully process all the data coming from user devices. In addition, with the development

V. Kovolenko · A. Muthanna (✉) · A. Koucheryavy
The Bonch-Bruevich Saint Petersburg State University of Telecommunications, Saint Petersburg, Russia
e-mail: muthanna.asa@spbgut.ru

A. A. Ateya
Department of Electronics and Communications Engineering, Zagazig University, Zagazig, Sharqia, Egypt

A. A. Abd El-Latif et al. (eds.), *Secure Edge and Fog Computing Enabled AI for IoT and Smart Cities*, EAI/Springer Innovations in Communication and Computing,
https://doi.org/10.1007/978-3-031-51097-7_10

of the "Internet of Things" concept, as well as 5G/IMT-2020 networks, new areas of application appear that have high requirements for data transmission delays. Therefore, for example, when organizing interaction in a network of unmanned vehicles, it may be necessary to provide a circular delay in the transmission of data packets at the level of 1 ms.

Thus, we can conclude that in order to meet the growing demand for providing ultralow data transmission latencies of the order of 1 ms and to reduce the load on the core network caused by traffic coming from user devices, it is necessary to organize systems based on new technologies that allow reduce transmission delays and optimize network load.

Such a system can be organized, for example, using unmanned aerial vehicles (UAVs), as well as the concepts of software-defined networks (SDNs).

However, systems using UAVs have one drawback: the UAV's built-in battery is limited. Therefore, over time, in such systems, it will be necessary to migrate services from UAVs with a discharged built-in battery to other UAVs in the network.

2 Software-Defined Networks

According to the Open Network Foundation, responsible for the development of this concept, SDN is a network infrastructure that is dynamically managed and capable of adapting, in which the control layer is separated from the data layer, which in turn provides the ability to programmatically control the network, as well as abstract the application layer and services from the level of network infrastructure [5, 6].

The year 2006 can be considered as the beginning of the SDN concept development because in this year, employees and specialists from Stanford and the University of California at Berkeley proposed this concept [5, 6]. The starting point for the SDN industry was the formation of the American company Nicira Networks, the founders of which were the authors of the SDN concept – professors Nick McKeon, Martin Casado, and Scott Schenker.

The main principles of software-defined networks can be considered [5, 6]:

- Separating the transfer process itself from the data flow control process.
- The presence of a program-controlled, unified, intelligent interface, independent of the supplier's equipment and placed logically between the data transmission levels and the control level.
- Management of the entire network as a whole, and not just its individual functional elements (in this case, management should be carried out using a specialized SDN controller).
- Virtualization of the network and its physical resources.

The main characteristics of software-defined networks can be presented in the following paragraphs [5, 6]:

- Centralized network management.
- Software configuration options.

- The direct control level programming due to the separation of the control level from the data level.
- Flexibility in regulating data transfer flows, provided by isolating the control level from the data transfer level.
- Independence of the network structure from equipment vendors and the use of open standards, which in turn simplifies the network architecture and its operation.

The main goal of creating software-defined networks is to separate the control plane from the data plane, which was proposed to be done by transferring the network management logic to a specialized device – the controller. This implementation approach provides a more convenient management of the network structure, which allows more efficient and faster implementation of new services.

Functionally, the SDN network architecture consists of three layers: the application layer, the control layer, and the data transfer layer.

1. Application layer.

   The application layer is responsible for implementing services.

   The interaction between the application and management layers is carried out through the north interface (northbound), which can be implemented through the organization of communication between the SDN controller and the application layer based on application programming interface (API).
2. Control layer.

   This layer implements the network management logic.

   In this case, the main functions of the control plane are carried out by the controller (network operating system) SDN. The north interface is used to interact with the application plane, and the southbound interface is used to interact with the data transfer plane. The most commonly used south interface is the OpenFlow protocol.
3. Infrastructure layer.

   At this level, there are switches that have the ability to support the protocol for interaction with the SDN controller (usually OpenFlow). Such devices can be, for example, OpenFlow switches.

   Within the SDN concept framework, a single server is considered as a centralized controller that manages data transfer flows between all switches of the infrastructure level. However, in large networks, using a single controller to control the flow of all switches can be cumbersome and undesirable. Therefore, it is most advantageous to divide a large corporate network into several SDN domains with their own SDN controllers.

   For interaction between SDN controllers or SDN controllers with other networks, the east/west (east/westbound) interface is used.

   The use of multiple SDN domains will ensure high scalability and reliability of the network architecture, the application of various privacy policies, as well as the phased deployment of network infrastructures.

Controllers can be placed in a certain area both densely and widely dispersed. In the first case, controllers allow for high throughput, which is most suitable for data centers. And in the second – for multi-position networks.

Traditionally, controllers are deployed horizontally, where a single network is split into several nonoverlapping subsets of data plane switches. Vertical deployment of SDN controllers is also possible, where management tasks are distributed among different controllers depending on criteria such as network type and location requirements.

3 Flying Software-Defined Network Architecture

In the previous section, software-defined networks were defined, and their main architecture layers and functional elements were discussed. Now it is proposed to consider the possibility of applying the concept of SDN in UAV systems.

The SDN concept due to such advantages as the use of open standards, the possibility of organizing a network infrastructure regardless of the equipment used (the flying segment is organized by UAVs from different vendors), the separation of the control plane from the data plane, as well as the implementation of a centralized and flexible network management, allows you to solve a number of the most pressing problems at the moment that arise in the organization of UAV systems [7]. Such problems include:

- The inconstant nature of communications between UAVs.
- The possibility of dividing the network into several separate fragments. The reason for network fragmentation can be the high mobility of the UAV.
- The most optimal use of UAV resources (since UAVs have limited computing resources and data storage capabilities).
- Restrictions on the possibilities of interaction with other UAVs and the ground segment (for example, with base stations, with sensory fields, or with user devices).
- Provide an overview of the network infrastructure globally (required for more efficient deployment of UAV systems).
- Increased network congestion and electricity consumption with a high demand for a certain type of traffic.
- The use of the SDN concept in UAV systems, in addition to providing the possibility of solving these problems, allows [7] the following:
- Implement centralized management of the network infrastructure. Centralized management, in turn, makes it possible to improve the quality of provided services and to use UAV computing, network, and storage resources more efficiently.
- Perform load balancing between UAVs and base stations in the most optimal way.
- Perform dynamic control of UAV movement in space.
- Reduce both network load and data packet loss.

- Provide the ability to reconfigure the network. Changing the network configuration can be done using a single SDN controller or using multiple controllers.
- Implement the flexible allocation of radio resources.

When applying the concept of SDN in UAV systems, the functions of SDN switches must be implemented on separate UAVs. The functions of the SDN controller can bc implemented in various ways both in the ground segment and in the air:

1. Implementation of SDN controller functions at ground base stations.
2. Implementation of SDN controller functions on stationary balloons. As part of the SDN concept, balloons act as an analog of base stations, but in the flying segment of the network.
3. Implementation of the functions of the SDN controller on the UAV.

Let us consider in more detail these methods of implementing an SDN controller in the case of a cluster organization of a flying segment of networks with UAVs:

- In the first and second ways of implementing the SDN controller, they can be used to manage flows between UAV clusters and other network devices (the interaction of UAVs of one cluster with UAVs of another cluster, with ground or air segment devices that are not included in UAV clusters, is carried out through the head node).
- The SDN controller in the UAV cluster can be implemented either only on the head node or on all UAVs of the cluster. In both cases, flow control is carried out only within one UAV cluster.
- When implementing the SDN controller only on the head node of the cluster, it is necessary to transfer the functions of the SDN controller from the head node to the UAV, which will be selected as the new head node of the cluster, when the internal battery of the old head node is discharged.
- When implementing the SDN controller on all UAVs of the cluster, the head node must operate in the SDN master controller mode, and all the others in the slave mode. When the headend's internal battery runs out, the headend's SDN controller switches to slave mode. At the same time, the SDN controller of the UAV selected as the new head node must switch to the master mode.

This method of implementing the SDN controller consumes less network resources of the cluster than the method with the implementation of the controller only on the head node; however, it uses more resources of individual UAVs and increases their power consumption (more functions performed – more power consumption).

- It is also worth noting that the functions of the SDN controller can be simultaneously implemented on ground base stations, fixed balloons and cluster UAVs. In this case, the controllers of base stations and balloons will be responsible for the interaction between UAV clusters and individual UAVs with other network devices, and the controllers on the UAV will be responsible for managing flows within the cluster.

4 Basic Architecture Elements

Before determining the main functions performed in UAV systems, it is necessary to determine the ways of organizing the network architecture and the possible scope provisions of such systems.

Networks in which UAVs are used as one of the main functional elements are wireless sensor networks that consist of two segments [8]:

- Terrestrial, which includes ground base stations, sensors, and various user devices. It is also worth noting that ground segment devices may either have the ability to self-organize network nodes (at the same time, interaction with these devices is carried out using the BLE, 6LoWPAN, and ZigBee protocols) or not have such an opportunity (interaction is carried out using the following protocols: RFID, Wi-Fi, and LoRa).
- Flying, organized by either single UAVs or their groups.

The use of UAVs is possible in a wide variety of areas of life. Consider several tasks for which UAVs can be used:

1. To perform the functions of base stations: In this case, a transceiver is installed on the UAV, which will be responsible for the information exchange with the user's mobile devices in the ground segment to establish communication between them. UAVs are used within the framework of this task to unload ground base stations (with a sufficiently high number of requests for establishing connections coming from mobile network users to ground base stations) or to organize communication in emergency situations when the ground infrastructure can be partially or completely destroyed.
2. To install sensors on the ground and then collect data from sensor fields: Due to high mobility, lower transmission delays, and higher air-to-ground bandwidth, UAVs collect data much faster than terrestrial base stations.
3. For data relay tasks: For these tasks, it is possible to store data and then transfer it to the user, in cases where data transfer using ground infrastructure is difficult or impossible (for example, if there is no ground infrastructure for communication with the user due to the geographical landscape of the area).
4. To fulfill certain user requests: This type of task, like the first one, is aimed at reducing the load on base stations.

Initially, only one or a small number of UAVs were enough to complete the tasks; however, the development of new technologies and an increase in the number of devices with network access will require the organization of systems with a large number of UAVs in the flying segment, divided into several different groups. There are several ways to organize UAV networks, among which centralized, cluster, and cellular architectures can be distinguished [9, 10].

In a centralized architecture, data exchange between several UAVs is carried out through the base station. Advantages: stability of the system when an individual UAV exits the operating state and the possibility of parallel execution of several tasks by individual UAVs. Disadvantages: proportional dependence of the network bandwidth on the UAV density and vulnerability of the ground infrastructure.

In cellular architecture, the ground segment is divided into several cells with their own frequency bands (to prevent interference from other cells), and UAVs can interact with each other both directly and using a base station. Emerging problems: the possibility of breakdown of base stations and the high cost of their deployment.

In the cluster organization of UAV networks, the flying segment is divided into several separate clusters. One of the UAVs of such clusters is selected as the head node. The head node is responsible for the data exchange between UAVs within the cluster and interaction with other clusters and the core network. The advantage of this architecture lies in the possibility of using UAVs with different communication and flight characteristics. Disadvantages: the inability of the UAV of one cluster to interact with the UAV of another cluster in case of partial or complete damage to the ground infrastructure.

5 The Main Functions Performed in the UAV Cluster

In the previous sections, four ways of using UAVs have been considered. Based on these methods and the possibility of applying the SDN concept in UAV systems, the list of main functions for UAVs will be as follows:

- SDN switch functions.
- Functions of the SDN controller: Of course, it is possible to implement controller functions only on ground base stations or stationary balloons, but in this case, both balloons and ground base stations may not be able to cope with the load coming to them.
- The base station functions: These functions include tasks related to collecting data from sensor fields, receiving user requests, or performing functions related to a communication channel between users of a cellular network.
- Base station controller functions: In this case, the UAVs perform functions similar to those of a cellular network base station controller.
- The head node functions (associated, for example, with the interaction of the UAV with devices outside the cluster).
- Functions for data relaying: These functions are related to data storage. Such features can be applied either to improve the quality of service on a discontinuous connection or to transmit data in delay-tolerant networks (for example, when transmitting data between two remote users through an area where there is no or poorly developed terrestrial network infrastructure).
- Functions for processing user requests: The easiest way to implement these functions will be to use all of the listed functions on each UAV. In this case, there would be no need to migrate services from one UAV to another when the internal battery runs out. However, the use of all functions on one UAV reduces the amount of free resources of both the UAV itself and the total resources of

the cluster (with the cluster organization of the flying network segment). At the same time, an increase in the number of UAV functions performed increases the amount of energy consumed by the UAV.

Therefore, it is more expedient to distribute functions among several UAVs. It is in this case that the migration of services from one UAV to another becomes the most urgent task.

6 The Proposed Algorithm

As mentioned earlier, the charge of the internal battery of the UAV is limited, and the implementation of all possible functions on each UAV is impractical since in this case the power consumption increases. In addition, the implementation of each service (function) on the UAV consumes a certain part of the UAV resources (computing, network, or data storage resources). Therefore, when the internal battery of the UAV is discharged, it will be necessary to migrate services from some UAVs (on which the charge of the internal battery has dropped below a certain level) to others (having sufficient resources to implement the transmitted services on them).

In this chapter, we propose an algorithm (sequence of actions) that is carried out when the internal battery of the UAV is discharged and includes the processes of determining the UAV for the subsequent migration of the service, the route of service migration, as well as the processes for sending the UAV to recharge the built-in battery and returning the UAV to the state of execution of tasks.

It should be noted that the processes of determining the UAV for subsequent migration, the migration route, as well as the nearest area for recharging the built-in UAV battery will be carried out on a specific computing device, which can be ground base stations, fixed balloons, or UAV cluster head nodes (if the flying network segment is divided into several clusters). At the same time, such computing devices must know information about the free resources of all UAVs under their control: located in their coverage area (for ground base stations and fixed balloons) or belonging to a specific cluster (for the head node of the UAV cluster). To do this, computing devices will periodically send requests for information about free resources for all UAVs under their control.

Figure 1 shows the process of UAV service migration, carried out under the control of the cluster head node, a fixed balloon, or a ground base station, as well as the processes of moving the UAV to the internal battery recharging zones and then returning (after full charging) to perform previously set tasks.

The proposed algorithm will consist of the following steps:

1. Sending the UAV when the charge of the internal battery drops to a certain level, a request to the computing device to which this UAV is subordinate, to determine the UAV for the subsequent migration of services, and migration routes and zones for recharging the UAV.

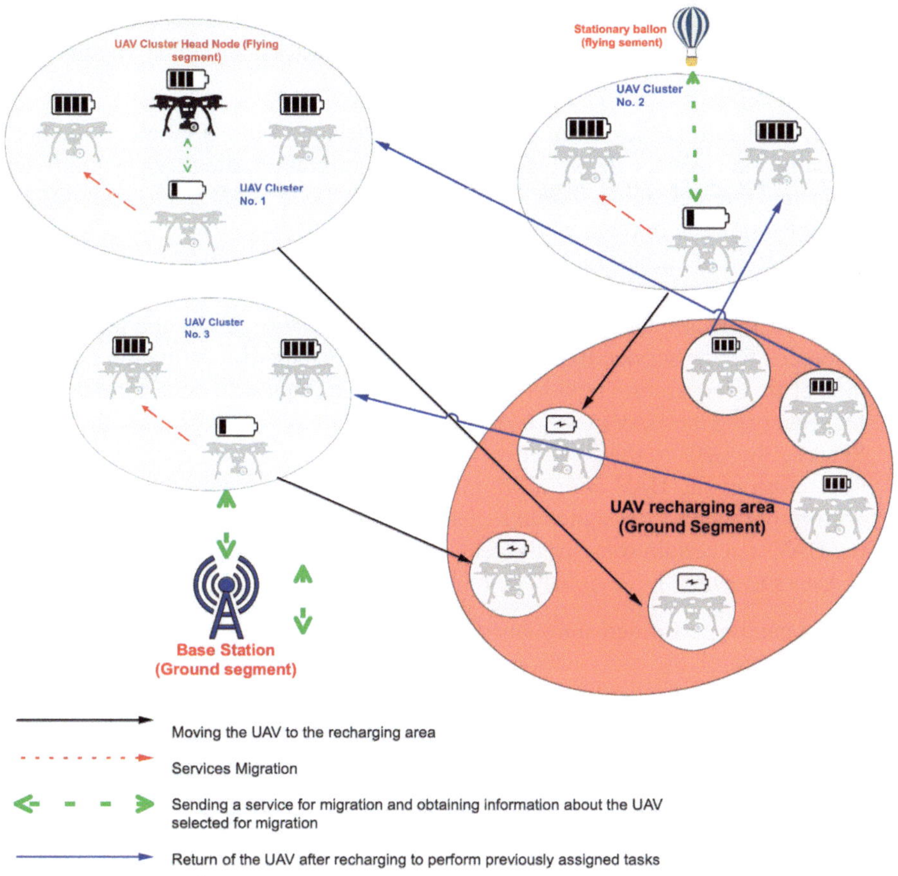

Fig. 1 Scheme of performing the service migration processes, sending the UAV for recharging, and returning to implementation of previously set tasks

2. Determination of the UAV by a computing device for the subsequent migration of services.
3. Determination of the migration route by the computing device.
4. Definition of a zone for recharging. This stage for the successful sending of the UAV for recharging should additionally include the following sub-stages:

    4.1 Definition of the charging zone: The main criteria for choosing a recharging zone should be the distance between the UAV and the recharging zone, calculated based on their coordinates in a three-dimensional space, as well as the availability of free landing places.

    4.2 Sending a request to the selected charging zone to reserve one place for UAV landing.

    4.3 Sending a response from the selected zone to the computing device about the possibility of recharging the UAV or sending a message about the absence of free places for UAV landing. Upon receiving a message about the lack of free places for landing, the process of choosing a zone for recharging the UAV is repeated.

5. Transfer of information from the computing device to the UAV about the definition of the UAV for the subsequent services migration, the services migration route, and the recharging zone.
6. Service migration implementation.
7. Confirmation of successful services migration: The UAV to which services need to be migrated must send information about either the migration completion or a request to re-implement the migration, or information about the lack of free resources for the service implementation. Upon receipt of information about the lack of free resources, steps 1 (without sending a request to determine the zone for recharging the UAV), 2, 3, and 7 are repeated.
8. Sending the UAV to the area for recharging.
9. The return of the UAV after a full charge of the internal battery to the state of performing the tasks assigned to it. This stage may consist in either the return of the UAV to a certain route or the return of the UAV to a certain cluster or group of UAVs to which it previously belonged.

Step 4 can be performed simultaneously with steps 2 and 3.

The scheme of the algorithm (sequence of actions) proposed in this chapter, which is carried out when the UAV's built-in battery is discharged, is shown in Fig. 2.

In some cases, it may not be possible to migrate all functions from a UAV with a discharged internal battery to one UAV: Some of the transferred functions may already be implemented on the UAV, or free resources of the UAV may be sufficient only to implement part of the functions. In this case, it would be more efficient to consider the migration of services and functions separately (the proposed algorithm should be performed separately for each type of service or for each separate group of functions considered earlier).

7 Conducted Experiment

As part of this work, three systems were compared in the MATLAB software package:

1. System (1) – UAVs without edge computing.
2. System (2) – UAVs with edge computing.
3. System (3) – UAVs with edge computing and service migration.

Figures 3 and 4 show how average latency and the number of blocked tasks, measured as a percentage, depend on the number of tasks performed per drone.

The histograms in Figs. 3 and 4 show that as the number of tasks performed by the drones increases, the average latency and the percentage of dropped tasks in

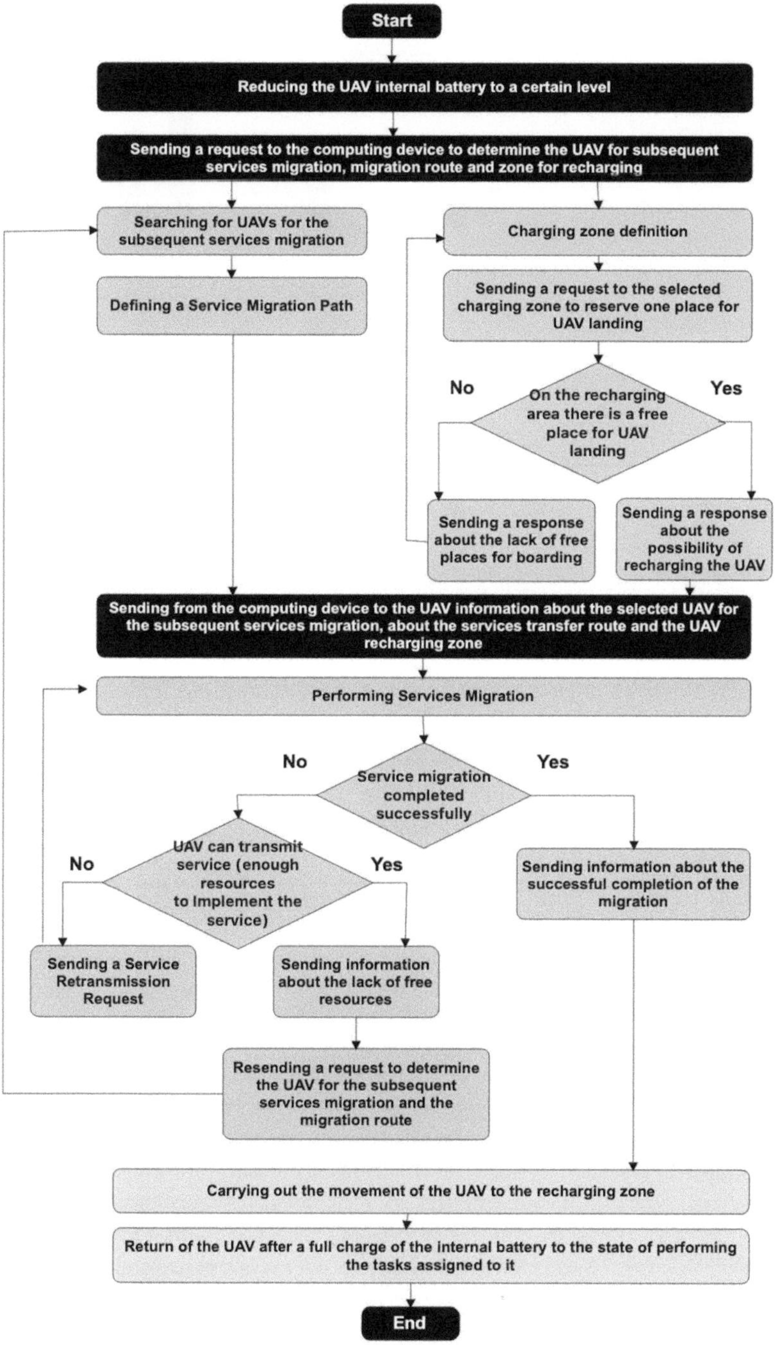

Fig. 2 The main stages of the proposed algorithm

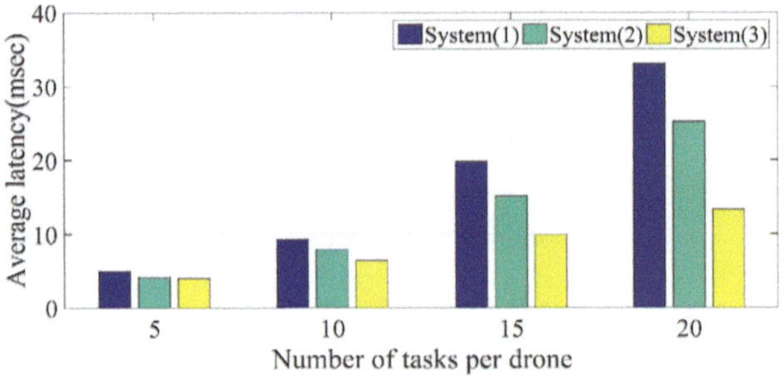

Fig. 3 The dependence of the average delay on the number of tasks performed by drones

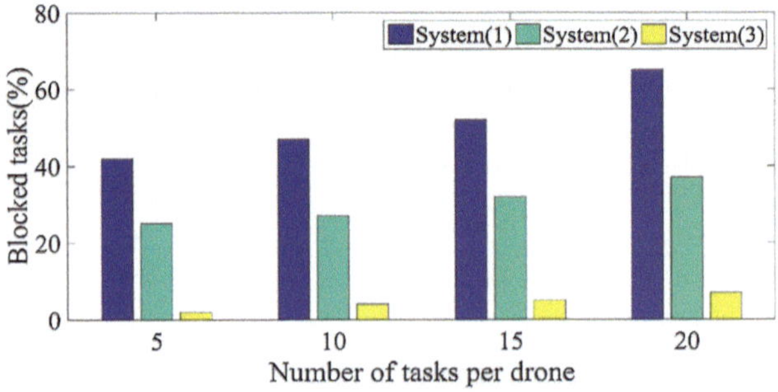

Fig. 4 The dependence of the number of blocked tasks (in percent) on the number of tasks performed by drones

system 3 (UAVs with edge computing and service migration) increase with a lower intensity compared to other systems under consideration.

Based on the experiment, it can be concluded that the use of UAV systems with service migration and edge computing makes it possible to achieve lower values of the average data transfer delay and the percentage of blocked tasks performed by UAVs.

8 Conclusion

In this chapter, possible ways to organize the architecture of networks with UAVs and the main areas of UAVs implementation were considered in detail. A description of the main benefits that UAV systems would receive from the implementation of the SDN concept, as well as several ways to implement SDN controllers, was given.

The chapter presented groups of functions necessary for the successful completion of all types of tasks (related to four areas of application) and for the implementation of the SDN concept in systems with UAVs. In addition, the need to migrate services from one UAV to another and why it is not profitable to implement all possible services and functions on each UAV were theoretically justified.

The chapter also proposed an algorithm (sequence of actions) carried out when the internal battery of the UAV is discharged and includes the processes of determining the UAV for the subsequent service migration, the service migration route, sending the UAV to the recharging zone and returning the UAV to perform assigned tasks.

In a comparison of the three systems conducted in the software package MATLAB, it was found that with an increase in the number of tasks performed per drone, the average delay and the number of discarded tasks (in percent) for system 3 (UAVs with edge computing and service migration) increased less intensively (compared to other systems under consideration: UAVs without edge computing and UAVs with edge computing).

Acknowledgments: The research was supported by Applied Scientific Research under the SPbSUT state assignment No. 1022040500653-0, 2023.

References

1. Kucheryavyj, A.E., Prokop'ev, A.V., Kucheryavyj, E.A.: Samoorganizuyushchiesya seti, 312 s. Tipografiya Lyubavich, SPb. (2011)
2. Gol'dshtejn B.S., Kucheryavyj A.E.: Seti svyazi POST-NGN, 160 s. BHV Peterburg, SPb. (2014)
3. Farhan, M., Jabbar, S., Aslam, M., Hammoudeh, M., Ahmad, M., Khalid, S.: IoT-based students interaction framework using attention-scoring assessment in eLearning M. Futur. Gener. Comput. Syst. **79**, 909–919
4. Ande, R., Adebisi, B., Hammoudeh, M., Saleem, J.: Internet of things: evolution and technologies from a security perspective. Sustain. Cities Soc. **54**, 101728. https://www.researchgate.net/publication/334724035_Internet_of_Things_Evolution_and_Technologies_from_a_Security_Perspective
5. Muhizi, S., Muthanna, A., Kirichek, R.V., Koucheryavy, A.E.: Analysis and modeling of load balancing in software-defined networks. Electrosvyaz. **1**, 23–29 (2019)
6. Ateya, A.A., Muthanna, A.S., Koucheryavy, A.E.: Intelligent core network for 5G and tactile internet systems based on software defined networks. Electrosvyaz. **3**, 34–40 (2019)
7. McCoy, J., Rawat, D.B.: Software-defined networking for unmanned aerial vehicular net working and security: a survey. Electronics. **8**(12), 1468 (2019)
8. Ateya, A.A., Muthanna, A., Kirichek, R., Hammoudeh, M., Koucheryavy, A.: Energy-and latency aware hybrid offloading algorithm for UAVS. IEEE Access. **T.7.C**, 37587–37600 (2019)
9. Katila, C.J., Okolo, B., Buratti, C., Verdone, R., Caire, G.: UAV-to-ground multi-hop communication using backpressure and FlashLinQBased algorithms. In: Proceedings of the IEEE 29th Annual International Symposium on Personal, Indoor and Mobile Radio Communications (PIMRC), pp. 1179–1184. IEEE (2018)
10. Nikolayevich, V., Alzaghir, A., Volkov, A., Muthanna, A., Koucheryavy, A.: Clustering algorithms for UAV placement in 5G and Beyond Networks: 2020. In: 12th International Congress on Ultra Modern Telecommunications and Control Systems and Workshops (ICUMT), pp. 301–307 (2020)

FedBA: Non-IID Federated Learning Framework in UAV Networks

Pei Li, Zhijun Liu, Luyi Chang, Jialiang Peng, and Yi Wu

1 Introduction

Due to the rapid development of network and communication technologies in recent years, drones have shown the great commercial value and are widely used in traffic monitoring, site management, aerial tourism photography and commercial performances [1]. Especially with the maturity and perfection of 5G technology and the continuous development of 6G technology, the computing speed as well as data transmission rate has been greatly improved [2]. Unmanned Aerial Vehicle (UAV) is gradually being used as an aerial service device with its excellent mobility, high computing power, and high perception capability [3]. Drones are uniquely suited for remote and dangerous areas (e.g., wildlife photo collection, atmospheric cloud weather prediction) as well as for scenarios that require extreme real-time performance (e.g., traffic management, mountain power maintenance, disaster prediction). In addition, the UAV cluster, as an aerial edge device, can collect data through its sensors and then transmit the data to the central server for processing and modeling [4]. However, this process can cause privacy leakage of users and increase communication costs [5].

The two main reasons for the above problems are as follows. First, when users upload their private data to a third-party central server if the third-party central server is attacked by hackers or the third-party central server is untrustworthy, it will cause the risk of user private data leakage. Second, traditional centralized machine learning requires that a large amount of data be uploaded to a third-party server

P. Li · Z. Liu · L. Chang · J. Peng · Y. Wu (✉)
School of Data Science and Technology, Heilongjiang University, Harbin, China
e-mail: 2212623@s.hlju.edu.cn; 2222708@s.hlju.edu.cn; 2202518@s.hlju.edu.cn; jialiangpeng@hlju.edu.cn; 1995050@hlju.edu.cn

A. A. Abd El-Latif et al. (eds.), *Secure Edge and Fog Computing Enabled AI for IoT and Smart Cities*, EAI/Springer Innovations in Communication and Computing,
https://doi.org/10.1007/978-3-031-51097-7_11

for computing. Due to the limited communication resources of UAVs, they cannot afford high communication costs [6].

In response to these challenges, Federated Learning (FL) [7] is gradually becoming the main focus of popular research. By merely uploading model parameters, federated learning addresses the issue of communication costs and lowers the danger of data leaking. Nevertheless, when multiple UAVs work jointly, the difference in the monitoring area of each UAV causes heterogeneity in the collected sensing data. This eventually causes the local device's ability to recognize data and the precision of the global model to decline [8, 9]. To this end, we propose a brand-new calculation method for an aggregate based on the distance between training models to palliate the non-IID problem caused by local data. The system overview of this paper is shown in Fig. 1. Consequently, the following are the contributions to this paper:

- We propose an FL-enabled framework for the Internet of drones that protects user privacy and reduces communication costs between drones.
- We propose a new FL aggregation algorithm named FedBA on non-IID data to cope with the statistical heterogeneity issue in the UAV-assisted FL framework. It is worth noting that this algorithm adds no additional computational cost to the drone.
- We assess our algorithm's viability for classification tasks on images from the real world on three datasets (i.e., CIFAR-100, MNIST, and Fashion-MNIST). From the final test results, it is found that our method has better test accuracy than the baseline and has a better advantage in convergence speed.

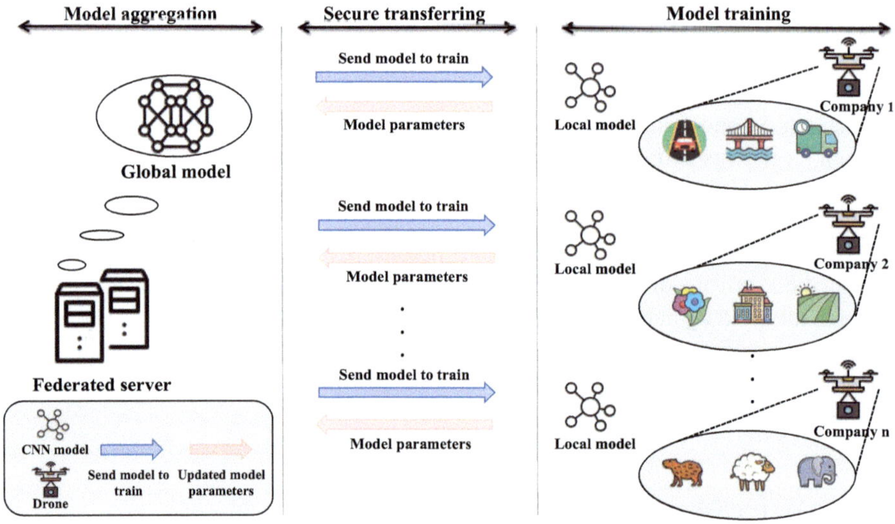

Fig. 1 There are three steps in the UAV-assisted federated learning framework approach. First, UAVs from various firms hover at low altitudes to gather pictures of objects in various locations. The UAV employs its computational capability to do local training on the data it has gathered. The drone then runs numerous local model training iterations until the model converges. The UAV then sends the model's parameters for aggregation to a dependable server in the third party to obtain a global model

2 Related Works

2.1 Privacy-Preserving UAV Image Recognition

In the last several years, UAVs have become the main means of acquiring data due to their advantages of multi-view image acquisition and extreme mobility [10]. For example, using drone image recognition technology for mountainous power grid inspection projects, construction material image recognition, rare animal identification detection, etc. However, in most cases, the images captured by drones are private data. This data cannot be easily transferred to the cloud for centralized training or shared [11]. Researchers have applied blockchain and federated learning frameworks to the task of drone image recognition, which allows for the privacy protection of image data captured by drones. For example, reference [12] used blockchain to authorize joint learning of disaster response systems using drones. Reference [13] suggested applying the FedProx [14] algorithm within the FL framework for usage in driverless traffic monitors to enhance the performance of driverless road transport.

2.2 Federated Learning on Non-IID Data

For federated learning, FedAvg [7] is the most representative algorithm. It can be applied to many different real-world scenarios, such as medical image model recognition, financial risk control systems, etc. However, non-IID data still hurts FedAvg. Solving the non-IID problem has also been a topic of great interest in federation learning [15]. For example, reference [16] created a common public dataset using data from local clients. Each client is pre-trained with a pre-prepared public dataset, and then the trained model is trained with its private data to mitigate the impact of data heterogeneity to a certain extent. Since client data are private, the researchers plan to reduce the discomfort brought on by the non-IID data by creating a unique model for each client as opposed to using the standard global model distributed by the server. Reference [17] adjusted the number of local iterations dynamically and introduces additional regularization terms. In this way, the gap between the server global model and the client local model is continuously narrowed, thereby continuously alleviating data heterogeneity. Reference [18] solved the statistical heterogeneity problem by adopting a personalized federation learning approach, which uses a personalization layer and a base layer in the model to achieve personalization, thus providing a unique model for each client to solve the statistical heterogeneity problem. Reference [19] created personalized models by local fine-tuning to solve non-IID problems.

3 Methodology

In the present paragraph, we describe the specific formula of federated learning as well as the training process. At the same time, we also describe the specific formula of the algorithm proposed in this article.

3.1 Federated Learning

As a distributed learning paradigm, federated learning has gradually entered the public's vision. It can satisfy multiple clients to realize data sharing based on protecting local data security, thereby solving the problem of data silos [20]. In the entire federated learning process, the transfer of gradients is used to replace the transfer of data, thereby ensuring the security of the data. At the same time, the client's private data is kept locally throughout the training process and will not be leaked to third-party servers. The goal of our final training is to minimize the global loss $\mathcal{L}(\cdot)$ until it converges, and the following equation is usually used as the objective function:

$$\mathcal{L}(\delta) = \sum_{k=1}^{K} p_k \mathcal{L}_k(\delta), \tag{1}$$

where $\mathcal{L}_k$ represents the k-th client's local loss, p_k is the relative contribution rate of the k-th client where $\sum_{k=1}^{K} p_k = 1$, and δ stands for model parameters. In the federated average (i.e., FedAvg [7]), the global loss function $\mathcal{L}(\delta)$ is as follows:

$$\mathcal{L}(\delta) = \sum_{k=1}^{K} \frac{n_k}{n} \mathcal{F}_k(\delta) \quad \text{where} \quad \mathcal{F}_k(\delta) = \frac{1}{n_k} \sum_{i=1}^{n_k} f_i(\delta), \tag{2}$$

where n represents all data, n_k is the amount of data owned by the k-th client, and $f_i(\delta)$ is the cross entropy loss function used to calculate the training loss of the i-th sample.

3.2 Proposed Algorithm

First, we define a distance function $\mathcal{A}(\|\delta^k - \delta\|^2)$. The purpose of the distance function $\mathcal{A}(\|\delta^k - \delta\|^2)$ is to evaluate the distance between two models, where the global model δ and the k-th local model δ^k. The function $\mathcal{A}(\cdot)$ has the following characteristics:

- $\mathcal{A}$ is nonlinear and an increasing convex function when $\|\delta^k - \delta\|^2 \in [0, \infty)$ with $\mathcal{A}(0) = 0$.
- When $\|\delta^k - \delta\|^2 \in [0, \infty)$, $\mathcal{A}$ can be differentiated continuously.

Specifically, a typical way to measure the nonlinear function of the distance between δ^k and δ is a negative exponential function with both maximum and minimum incentive penalty functions [21]. In this paper, we use the following method to measure the distance between δ^k and δ.

$$\mathcal{A}(\|\delta^k - \delta\|^2) = \log[g(\|\delta^k - \delta\|^2)]. \tag{3}$$

Here, we define $g(\cdot) \in [0, \pi/2)$ and the $g(\cdot)$ function formula is as follows:

$$g(x) = \begin{cases} x, & 0 \le x \le 1 \\ \arctan(x), & \text{otherwise} \end{cases}. \tag{4}$$

Next, we present the FedBA algorithm proposed in this paper. Suppose there are K clients in the federated learning training. The $\mathcal{D}_k = \{x_i, y_i\}_{i=1}^n$ denotes the local private dataset of client k, where x_i is the input data sample (e.g., images captured by the UAVs), and y_i is the corresponding target output (e.g., recognized category labels). The FedBA algorithm's precise procedure is as follows.

- ***Step 1, Local Model Training:*** First, in round t, each local client receives the aggregated global model from the server side. Second, each local client uses its private dataset to iteratively update the model parameters using Stochastic Gradient Descent (SGD) method. The formula for completing the update for the k-th client model is as follows:

$$\delta_t^k \leftarrow \delta_t - \eta \nabla l(k, \delta_t, \mathcal{D}_k, e, b), \tag{5}$$

  where η is the gradient descent algorithm's learning rate, δ_t^k is the model after the k-th client completes the SGD update in t communication rounds, δ_t is the global model issued by the server in completing t communication rounds, and $l(\cdot)$ represents the cross entropy loss function.
- ***Step 2, Aggregation:*** The server computes the Euclidean distance between each local model and the global model of the previous communication round. Because the range of Euclidean distance values between models spans too large a range and this Euclidean distance function is not a finite function in terms of derivation itself. So we transform the obtained Euclidean distance using the $\mathcal{A}(\cdot)$ function and then use the obtained values to calculate the aggregation weights to obtain the global model for the new round. The calculation formula is as follows:

$$p_{t+1}^k = \frac{\mathcal{A}_k(\|\delta_{t+1}^k - \delta_t\|^2)}{\sum_{k \in G_t} \mathcal{A}_k(\|\delta_{t+1}^k - \delta_t\|^2)}, \tag{6}$$

where $\mathcal{A}$ is a function of the distance measured between δ_{t+1}^k and δ_t; G_t is the collection of round t's participating training clients. The server calculates the model parameters received in this communication and obtains the aggregated weight p_{t+1}^k of each client. Afterward, the server aggregates the local models of the client according to the calculated weights to obtain a new global model δ_{t+1}. The formula is as follows:

$$\delta_{t+1} = \sum\nolimits_{k\in G_t} p_{t+1}^k \delta_{t+1}^k. \tag{7}$$

- ***Step 3, Broadcast Global Model:*** The server broadcasts the next round of global model δ_{t+1} to each client to continue training.

Algorithm 1 describes all the processes in which FedBA continuously iterates between the server and the client.

Algorithm 1: FedBA

Input: The number of clients is K, the sampling rate of the client is C, the mini-batch size $\mathcal{B}$ and the number of updates of the local model is E, and the learning rate during the SGD of the model is η.
Output: Training to achieve convergence of the global model δ^*.
On the server side:
Initialize a model parameter δ_0
for *iteration round* $t = 1, 2, \ldots$ **do**
  $G_t \leftarrow random\ set\ of\ m = max(C \cdot K, 1)\ clients$
  for *each client* $k \in G_t$ *in paraller* **do**
    $\delta_{t+1}^k \leftarrow \delta_t^k$
    $p_{t+1}^k = \frac{\mathcal{A}_k(\|\delta_{t+1}^k - \delta_t\|^2)}{\sum_{k\in G_t} \mathcal{A}_k(\|\delta_{t+1}^k - \delta_t\|^2)}$
    $\delta_{t+1} = \sum_{k\in G_t} p_{t+1}^k \delta_{t+1}^k$
  end
  Broadcast δ_{t+1} to each client
end
In the k-th client:
for *local training iteration round* e $\leftarrow$ 1 *to* E **do**
  for *mini-batch size* $b \in \mathcal{B}$ **do**
    $\delta_t^k \leftarrow \delta_t - \eta \nabla l(k, \delta_t, \mathcal{D}_k, e, b)$
  end
  Return δ_t^k to server
end

---

4 Experiment

In this subsection, we assess the effectiveness of our proposed framework and contrast it with FedAvg [7]. Our experimental environment is configured with Python version 3.9.12 and PyTorch version 1.12.1.

4.1 Experiment Setup

We use three real-world datasets CIFAR-100,[1] MNIST,[2] and Fashion-MNIST[3] to simulate the images collected during drone operations. The description of the dataset is as follows: the CIFAR-100 dataset contains 100 common types of animals and objects in life and each category contains 600 pictures and 10,000 pictures are selected as the test set. During the training process, each of our clients was assigned 2500 images as a training set. The MNIST dataset contains a total of 70,000 digit pictures of ten categories in various handwritten versions and includes 10,000 pictures in the test set. Also during the training process, we allocated 3000 images to each client as a training set. For the Fashion-MNIST dataset, we allocated 3000 training data to each customer during the entire training process and the overall dataset, contains a total of 70,000 data, and ten categories belong to the category of clothes. At the same time, we adopted the most mainstream non-IID division method, that is, Dirichlet segmentation. We use the Dirichlet distribution function to split the dataset, and divide the dataset to all local clients at one time. To achieve the simulation of data heterogeneity UAVs collect pictures in different areas. By modifying the Dirichlet distribution function's parameter (i.e., μ), it is feasible to influence the degree of non-IID data to mimic the distribution of real-world data. Therefore, we adjust the parameter μ so that the sample labels are distributed differently on each client. More precisely, the samples of each category label in our experimental data were distributed among different clients according to different proportions so that each client contained all category labels, but with different amounts of label data.

4.2 Training Details

In the experimental process of this paper, we use SGD to update a six-layer (convolutional layer, pooling layer, and fully connected layer each occupying 2

[1] http://www.cs.toronto.edu/~kriz/cifar.html

[2] http://yann.lecun.com/exdb/MNIST

[3] https://github.com/zalandoresearch/Fashion-MNIST

layers) Convolutional Neural Network (CNN) model. In addition, we selected $K = 20$ as the maximum number of clients, the sampling rate of clients $C = 0.6$, the learning rate in gradient descent $\eta = 1e - 3$, the number of local epochs $E = 5$, the mini-batch size $\mathcal{B} = 64$, and the parameters of the Dirichlet distribution function $\mu = 1e - 1$.

4.3 Experiment Result

We evaluate the FedBA algorithm on the non-IID settings with the CIFAR-100, Fashion-MNIST, and MNIST datasets. The main advantage of our experiment is mainly reflected in the better test accuracy achieved in the experiment, and at the same time, the experimental results show that our algorithm has a better convergence speed than the baseline, which is more important in the field of UAVs. The analysis of accuracy is detailed below.

- As shown in Fig. 2, we evaluated the accuracy of FedBA in the CIFAR-100 dataset non-IID scenario. With the increasing number of communication rounds, the performance of our model continues to improve and the accuracy rate reaches 34.29%, and the baseline final accuracy only converges to 31.07%. Compared to the baseline FedAvg, the accuracy of FedBA was improved by 3.22%. This is caused by the fact that the baseline FedAvg uses only the simple idea of average aggregation, which does not take into account the different weights assigned to each client in the process of average aggregation.
- Figure 3 shows the accuracy of FedBA in the non-IID scenario of the MNIST dataset. When the final training model converges, we can see that FedBA improves by 0.85% compared to FedAvg. The result made a small contribution

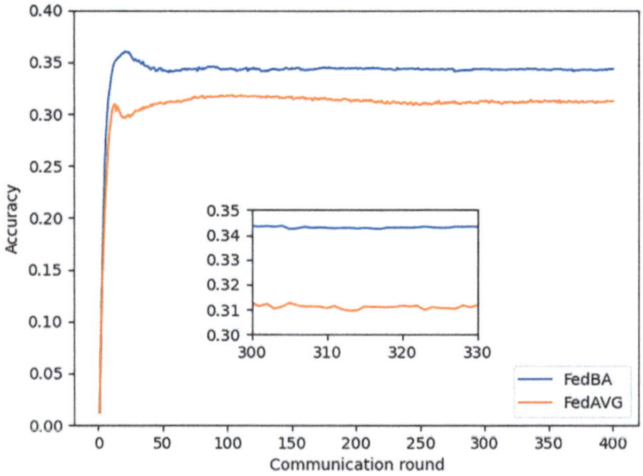

Fig. 2 Test accuracy curve on CIFAR-100 dataset

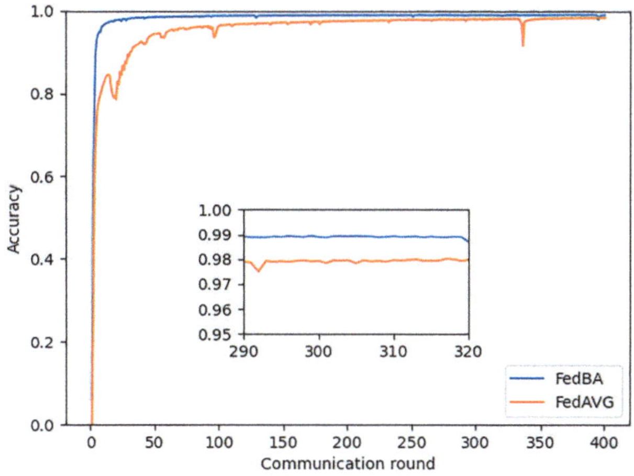

Fig. 3 Test accuracy curve on MNIST dataset

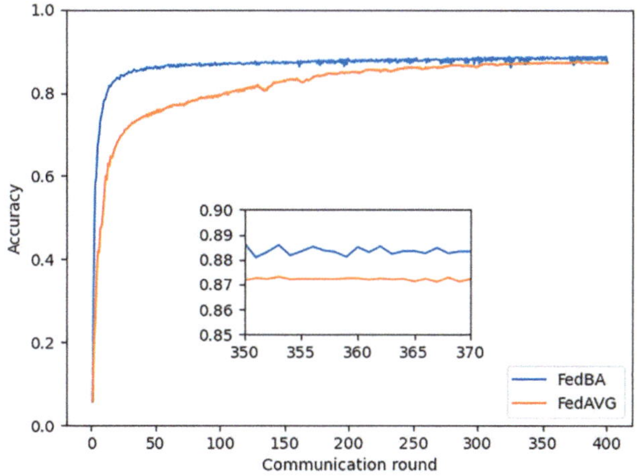

Fig. 4 Test accuracy curve on Fashion-MNIST dataset

to the improvement, this is because the MNIST dataset is much simpler and the accuracy is very high. It is difficult to make a large improvement.

- As shown in Fig. 4, we evaluated the accuracy of FedBA in the Fashion-MNIST dataset non-IID scenario. The accuracy of our algorithm FedBA is higher than the baseline FedAvg and has a faster convergence rate throughout the 500 communication rounds. Specifically, FedBA achieves 88.86% accuracy at final convergence, yet FedAvg only achieves 87.17%.

5 Conclusion

In this paper, we propose an FL-enabled framework for the Internet of drones that protects user privacy and reduces communication costs between drones. At the same time, we modify the aggregation method in federated learning to obtain better accuracy and convergence speed from test experiments. In the improvement of the whole method, we have taken into account the non-independent and identical distribution problem caused by the UAV collecting data from different areas. Therefore, it is reasonable to assume that different local UAV models should be assigned different aggregation weights in the federated learning aggregation phase to obtain a higher-quality global model. Overall, our method has achieved good experimental results, but we can continue to conduct further research in the direction of model encryption in the future.

Acknowledgments The Heilongjiang Provincial Natural Science Foundation of China (Grant No. LH2020F044), the 2019-"Chunhui Plan" Cooperative Scientific Research Project of the Ministry of Education of China (Grant No. HLJ2019015), and the Fundamental Research Funds for Heilongjiang University of China (Grant No. 2020-KYYWF-1014) have all contributed to the funding of this work.

References

1. Kouhdaragh, V., Verde, F., Gelli, G., Abouei, J.: On the application of machine learning to the design of uav-based 5g radio access networks. Electronics **9**(4), 689 (2020)
2. Chowdhury, M.Z., Shahjalal, M., Ahmed, S., Jang, Y.M.: 6g wireless communication systems: Applications, requirements, technologies, challenges, and research directions. IEEE Open J. Commun. Soc. **1**, 957–975 (2020)
3. Zhang, H., Hanzo, L.: Federated learning assisted multi-uav networks. IEEE Trans. Vehic. Technol. **69**(11), 14104–14109 (2020)
4. Bithas, P.S., Michailidis, E.T., Nomikos, N., Vouyioukas, D., Kanatas, A.G.: A survey on machine-learning techniques for uav-based communications. Sensors **19**(23), 5170 (2019)
5. Gupta, L., Jain, R., Vaszkun, G.: Survey of important issues in uav communication networks. IEEE Commun. Surv. Tutor. **18**(2), 1123–1152 (2015)
6. Lee, B.H.-Y., Morrison, J.R., Sharma, R.: Multi-UAV control testbed for persistent uav presence: Ros gps waypoint tracking package and centralized task allocation capability. In: 2017 International Conference on Unmanned Aircraft Systems (ICUAS), pp. 1742–1750. IEEE, Piscataway (2017)
7. McMahan, B., Moore, E., Ramage, D., Hampson, S., y Arcas, B.A.: Communication-efficient learning of deep networks from decentralized data. In: Artificial Intelligence and Statistics, pp. 1273–1282. PMLR (2017)
8. Sattler, F., Wiedemann, S., Müller, K.-R., Samek, W.: Robust and communication-efficient federated learning from non-iid data. IEEE Trans. Neural Netw. Learn. Syst. **31**(9), 3400–3413 (2019)
9. Zhu, H., Xu, J., Liu, S., Jin, Y.: Federated learning on non-iid data: a survey. Neurocomputing **465**, 371–390 (2021)
10. Oleksyn, S., Tosetto, L., Raoult, V., Joyce, K.E., Williamson, J.E.: Going batty: the challenges and opportunities of using drones to monitor the behaviour and habitat use of rays. Drones **5**(1), 12 (2021)

11. Yao, J., Ansari, N.: Secure federated learning by power control for internet of drones. IEEE Trans. Cognit. Commun. Netw. **7**(4), 1021–1031 (2021)
12. Pokhrel, S.R.: Federated learning meets blockchain at 6g edge: A drone-assisted networking for disaster response. In: Proceedings of the 2nd ACM MobiCom Workshop on Drone Assisted Wireless Communications for 5G and Beyond, pp. 49–54 (2020)
13. Donevski, I., Nielsen, J.J., Popovski, P.: On addressing heterogeneity in federated learning for autonomous vehicles connected to a drone orchestrator (2021). Preprint arXiv:2108.02712
14. Sahu, A.K., Li, T., Sanjabi, M., Zaheer, M., Talwalkar, A., Smith, V.: On the convergence of federated optimization in heterogeneous networks (2018). Preprint arXiv:1812.06127, 3:3
15. Li, Q., Diao, Y., Chen, Q., He, B.: Federated learning on non-iid data silos: An experimental study. In: 2022 IEEE 38th International Conference on Data Engineering (ICDE), pp. 965–978. IEEE, Piscataway (2022)
16. Zhao, Y., Li, M., Lai, L., Suda, N., Civin, D., Chandra, V.: Federated learning with non-iid data (2018). Preprint arXiv:1806.00582
17. Li, T., Sahu, A.K., Zaheer, M., Sanjabi, M., Talwalkar, A., Smith, V.: Federated optimization in heterogeneous networks. Proc. Mach. Learn. Syst. **2**, 429–450 (2020)
18. Arivazhagan, M.G., Aggarwal, V., Singh, A.K., Choudhary, S.: Federated learning with personalization layers (2019). Preprint arXiv:1912.00818
19. Deng, Y., Kamani, M.M., Mahdavi, M.: Adaptive personalized federated learning (2020). Preprint arXiv:2003.13461
20. Chu, L., Wang, L., Dong, Y., Pei, J., Zhou, Z., Zhang, Y.: Fedfair: Training fair models in cross-silo federated learning (2021). Preprint arXiv:2109.05662
21. Huang, Y., Chu, L., Zhou, Z., Wang, L., Liu, J., Pei, J., Zhang, Y.: Personalized cross-silo federated learning on non-iid data. In: AAAI, pp. 7865–7873 (2021)

Part II
Fog/Edge Computing Security Issues

Big Data Analytics for Secure Edge-Based Manufacturing Internet of Things (MIoT)

Lo'ai Tawalbeh, Swathi Lakkineni, Fadi Muheidat, Ummugul Bulut, and Ahmed A. Abd El-latif

1 Introduction

Modern big data analytics systems allow for quick and efficient analytical procedures. In the digital world, organizations grow by continually scanning and analyzing data. However, today, such methods have become increasingly challenging with many organizations as they lack formal processes and systems to monitor these practices properly. This is where machine learning and deep learning come in. In machine learning, a machine attempts to identify a meaningful feature that identifies data in text or pictures. It is in this sense that deep learning understanding comes to us as a descriptive term. However, it cannot be used in the marketing context to describe what is going on. Even with deep learning, there is still a question of translating the meaning of deep learning into actionable insight. The

L. Tawalbeh (✉)
Computer Engineering Department, Jordan University of Science and Technology, Irbid, Jordan
e-mail: tawalbeh@just.edu.jo

S. Lakkineni
University of the Cumberlands, Williamsburg, KY, USA

F. Muheidat
California State University San Bernardino, San Bernardino, CA, USA

U. Bulut
University of the Incarnate Word, San Antonio, TX, USA

A. A. Abd El-latif
Center of Excellence in Quantum and Intelligent Computing, Prince Sultan University, Riyadh, Saudi Arabia

Department of Mathematics and Computer Science, Faculty of Science, Menoufia University, Shebin El-Koom, Egypt
e-mail: aabdellatif@psu.edu.sa

A. A. Abd El-Latif et al. (eds.), *Secure Edge and Fog Computing Enabled AI for IoT and Smart Cities*, EAI/Springer Innovations in Communication and Computing,
https://doi.org/10.1007/978-3-031-51097-7_12

second central point is that deep understanding is still not a general-purpose learning technique and has many specific challenges.

Organizations started to understand the importance of big data and realized that Internet of Things (IoT) devices are significant sources of collecting big data. Thus, the results of big data analytics help in decision-making (Verma n.d.). Smart city and smart grid technologies enable smart homes and cities to work together with conventional centralized infrastructures, allowing IoT devices to communicate with each other (Lavalle et al. 2020). IoT devices also connect to databases, which facilitate the delivery of intelligent services. The emergence of intelligent cloud systems and cloud-based applications enables the creation of intelligent services and environments on demand. Smart cities and intelligent grids allow information and communication services to be made available through wireless networks throughout our towns and cities. But these services cannot be provided without the use of wireless networks, which are continually being upgraded and improved (Lavalle et al. 2020).

Additionally, big data analytics provides various advantages—it can be used for better decision-making preventing fraudulent activities, among other things. This section explores a few essential concepts, briefly summarizes the data science field, and introduces a few of big data analytics's critical drivers. The area of data science has grown tremendously over the past decade. We are no longer the data scientists who created data but rather data-driven decision-makers. Data science uses machine learning to process and organize the data into meaningful, actionable knowledge (Chhabra et al. 2020). It provides an added value by identifying trends and identifying areas that warrant further investigation or enhancement. The analysis stage of the data analysis process is concerned with determining whether the information relating to a particular customer is useful for decision-making and strategy (Chhabra et al. 2020). Data analytics can further be applied to IoT devices.

IoT applications are being developed to help companies achieve both productivity and profitability [1]. Companies need to have a better understanding of their customers' needs. If a company's business is not sustainable, it can lose control over its operations, thereby hurting the organization and even the world. The most important aspect of this is that these are "internal" database resources such that most queries only require changes to internal data (Chhabra et al. 2020). These exciting trends also promote the development of intelligent systems for industrial, commercial, agricultural, and other industrial applications. However, such smart systems' potential is also of great concern due to the need to ensure reliable operation and maintain industrial systems availability. One of the current problems with traditional manufacturing is the lack of accurate and timely information on parts and component selection, manufacturing and quality control, and other operational aspects (Chhabra et al. 2020). Because of it, companies like Honeywell, Intel, and Motorola have moved to develop their products. One of the critical ingredients of advanced manufacturing technology is control, monitoring, and training. Management is vital here, as well. The idea is to control the environment in which you are operating. In a factory, for example, employees are monitored continuously. The entire process must be kept under the control of management (Chhabra et al. 2020).

IoT is essentially a platform where embedded devices are connected to the Internet to collect and exchange data with each other (Chhabra et al. 2020). There are currently over 100 such IoT devices available on the Internet. These devices contain and exchange data through wireless network protocols, such as Wireless Local Area Network (WLAN), cellular, and the Internet (Chhabra et al. 2020). Two devices connected with a wireless network, for example, two laptops connected in such a way that only one laptop connects to an external network, relates to a personal wireless network, such as a public wireless network. A wireless network connection can be thought of as an extension of the human body. In this sense, wireless networking does not even exist. As mentioned above, many wireless technologies are currently being developed to provide wireless connectivity in our homes and businesses (Chhabra et al. 2020).

Internet of Things slowly rose from consumer usage to usage in the manufacturing process. Because of IoT's benefits (Internet of Things) in the manufacturing process, industry owners are fostered to implement them. Although the generated data from manufacturing Internet of Things (MIoT) adds value to the industrial process, there are challenges encountered in analyzing the collected and stored data.

2 Data Analytics in Manufacturing

Increased demands can make the manufacturing process complex. It is one reason why the users fail to understand an overview of the system status because the system is distributed over different computer systems. The data generated in manufacturing is vast, and users get overloaded concerning fault detection, process monitoring, and data analysis. It results in late detection of failures, and sometimes the maintenance intervals are not chosen wisely. Issues related to process monitoring, anomaly detection, and data acquisition need to be addressed.

Data acquisition is the process of filtering, gathering, and cleansing data. In the manufacturing industry, the data is stored in a distributed fashion without any definition of semantics and time synchronization. Process monitoring and anomaly detection are the processes of detecting faults in the systems through continuous monitoring. For problems that relate to data integration, time synchronization and algorithmic solutions do exist.

3 Data Analytics

The science of analyzing raw data to find faults and patterns and draw conclusions is related to the information. Data analysis is a general term that is most commonly used to describe voluminous data, presenting computational and data-handling challenges. Data analytics relates to big data and every process that looks at historical data [2]. This data keeps growing exponentially over time, and the systems that analyze these data have also been evolving. Business data is now used to derive information through analytics. Today, most of the database administrators have generated reports on data related to the business.

Data can transform a business from predictive decision-making to prescriptive decisions. Big data can be helpful when paired with big analytics. Big data can be applied in every industry, such as finance, healthcare, retail, and manufacturing. It gives administrators a big picture of what is happening within the organization [11]. Manufacturing companies deploy a wide variety of applications that generate large amounts of data. Manufacturing applications such as Computerized Maintenance Management System (CMMS), Enterprise Resource Planning (ERP), and Manufacturing Execution System (MES), when an administrator integrates these applications with big data, patterns can be found. These patterns indicate how the factory floor is running. It quickly identifies the problem and mitigates the risk of a line-down situation at the manufacturing site.

The primary idea behind big data is that it provides a bigger picture of the generated data. Big data combines the usage of sensors, design data, etc. to identify the data pattern and convert it into a meaningful decision. In manufacturing, big data can include the data generated from electrical machines, sensors, and also the data produced by third-party vendors. Any data that is generated can feed into big data. Manufacturing generates voluminous data, and without analyzing the data, it would be impossible to understand the data trends.

Data visualization needs to be leveraged to gain insight into the data collected [7]. The process of data analysis helps the organization in decision-making. It helps them improve the manufacturing process, manage the supply chain, evaluate potential risks, customize product design, and ensure product quality. Manufacturing continues to remain important for the world economy. The role it plays in developing economies has dramatically shifted and continues to drive productivity, innovation, and efficiency gains. Big data is driving innovations by uncovering new insights.

4 Data Acquisition

Data generated by heterogeneous sources such as process data, ERP data, process data, or MES data must be analyzed in a central infrastructure [5]. Therefore, data acquisition needs to cope with products from different vendors, proprietary protocols, time synchronization of data, and data integration with standard interfaces. The data can be acquired at any layer of the automation pyramid. At the field or control level, data consists of Transmission Control Protocol/Internet Protocol (TCP/IP) or Structured Query Language (SQL) blocks. Data loggers can be deployed to extract the information from sensors and actuators. In manufacturing with distributed systems, Supervisory Control and Data Acquisition (SCADA) can provide insight into the type of data or equipment logs through a graphical user interface. The combination of various methods and proprietary protocols leads to heterogeneous networks. Today, no unified data acquisition method allows for the universal application of data mining methods.

5 Anomaly Detection

Anomaly detection can be done either using a model-based approach or a data-driven approach. The model-based approach uses a model that will demonstrate manufacturing under normal behavior. If the measurements from the simulation result vary significantly, the behavior is classified as abnormal. Neural network and regression-based methods have been widely used to determine the functional dependency between process variables. One of the statistical approaches to anomaly detection is building a probability distribution model in which state variables are employed to understand the temporal transition of process variables hidden. Kalman filters and particle filters are examples of such approaches. At the same time, data-driven anomaly detection has been used by many in recent years. This model depends on the measurable process data and does not require an analytical model. Principal component analysis (PCA), sub-space aided approach (SAP), and partial least squares (PLS) are a few examples of such approaches. Data-driven anomaly detection is capable of detecting anomalies that were not seen through a model-based approach. Histogram-based Outlier Score (HBOS), Fast Local Outlier Factor (FastLOF), and Local Distance-based Co-occurrence Outlier Factor (LDCOF) are a few algorithms developed for detecting anomalies in large-scale data sets.

Many companies such as oil and gas, chemical producers, car manufacturing, and refineries are examples of manufacturing organizations that use big data. These organizations also deploy sensors, dashboards, and scoreboards to analyze data. However, as the company's size keeps increasing, data increases and big data analytics can bring a monumental difference in its growth. A new dimension added through IoT brings the way assets and sensors relay information in real time. This data generated by IoT devices is of extreme value to the manufacturers, but many are overwhelmed with a large amount of data. Data analytics can help administrators sort out the data, cleanse, and analyze the machine data. Most companies understand the importance of big data analytics and started making investments in integrating data and managing assets to achieve a competitive edge.

The era of big data has just emerged. Big data can be a critical tool, particularly in the manufacturing sector, where process complexity, capacity restraints, and process variability are involved. Most manufacturing business leaders have now realized that big data analytics is not a "nice–to-have" option but a mandatory process needed to achieve efficiency and gain insight into the type of data generated in an increasingly complex environment. The organization that is looking to achieve a competitive edge should look to adopt big data analytics that seamlessly integrates with the business data and manage them. When evaluating big data analytics, business leaders need to consider the capabilities that align with the type of sector and helps them with supply chain, inventory management, optimized forecasting, stock replenishment, and other critical functions. Integrating business with the right data management platform allows organizations to leverage their strategic values, increase profits, improve operations, and improve relationships with customers and suppliers.

6 Data Analytics in MIoT

A fundamental change is occurring in the automotive sector according to the automated production market to intelligent manufacturing. In this development, the Internet of Things (IoT) plays the critical task of relating the physical world to cyberspace. Manufacturing of IoT (MIoT) comprises various production devices, sensors, electronic controls, controllers, Radio Frequency Identification (RFID) tags, and smart meters attached to software applications via computer platforms. Links for wire and wireless information exchange. There is an upswing in a considerable amount of MIoT-generated data traffic.

7 Characteristics of MIoT

Designers classify IoT into customer IoT, Internet of Things (CIoT), and manufacturing of Internet of Things (MIoT). MIoT is contrasted with CIoT in Table 1. By reference, CIoT primarily serves customers with MIoT. CIoT consists of portable devices (e.g., cell phones and electronics for portable use) and smart appliances (e.g., refrigerators, TVs, and machines for washing). CIoT is specifically designed to enhance users.

Although MIoT focuses primarily on improving the factory experience, transactions, and manufacturing, reducing the outage of equipment improves the consistency of goods. MIoT programs are usually spacecraft and are responsive to devise malfunction or downtime of machinery.

8 Comparative Analysis: MIoT vs. CIoT

Table 1 presents a comparison between MIoT (Industrial Internet of Things) and CIoT (Consumer Internet of Things) based on several key aspects. MIoT primarily targets industrial applications, focusing on devices such as sensors, smart meters, and actuators. In contrast, CIoT caters to consumer needs with smart devices

Table 1 Comparison between MIoT and CIoT

	MIoT	CIoT
Goal	Industry	Consumer
Devices	Sensors, smart meters, actuators	Consumer smart devices
Data rate	High	Average
Mission	Critical	Not critical
Delay	Sensitive	Can be tolerated
Working environment	(Noisy, temperature: High and low	Moderate

designed for personal use. MIoT devices typically operate at high data rates, whereas CIoT devices typically have an average data rate. MIoT emphasizes mission-critical operations, while CIoT functions are generally not critical. Additionally, MIoT systems are sensitive to delays, whereas delays in CIoT systems are more tolerable. Finally, MIoT devices are designed to function in challenging environments characterized by noise and extreme temperatures, whereas CIoT devices typically operate in more moderate conditions.

8.1 Data Acquisition

Techniques used for data analysis comprise transmission. Next, the processing of data includes procurement via specialized data collection techniques, and raw data from different sources throughout the entire development process RFID tags, for example, are reviewed by RFID, service warehouse observers. Then, the data that was obtained via the data collection system will be transferred, particularly wired networking networks or wireless systems—specifications about supporting data collection technology [17].

8.2 Data Processing and Storage

After data gathering, until maintaining the raw data, the raw data must be trained and tested. Owing to the significant number of data storage facilities and the inherent confusion regarding the characteristics of raw data, preprocessing methods in the data phase typically involve data cleaning, integration, and consolidation to ensure accurate verification. Storing details refers to the enormous process of holding and handling sets of results. The data management system is divided into two main components: infrastructure for computing and data management tools. The infrastructure includes storage, applications, and network devices that connect to the storage tools of the devices. Besides the networked ones, data processing software and storage systems are also available for the database management system.

8.3 Data Analytics

Various sophisticated management systems are used in the data analysis process to collect helpful information. From the large data sets for output, approximately, we characterize four kinds of information analytical systems:

I. Mathematical modeling.
II. Visualization of computer.
III. Data modeling, mining.
IV. Learning by computers. Info on allowing data processing technologies.

For instance, mathematical modeling of the spread of malware is useful for identifying possible weaknesses and strength of malicious software, so determine the feasible defense lines [8]. The behavior of malware spread is usually modeled by using the traditional epidemic models such as susceptible-infection (SI), susceptible-infection-susceptible (SIS), and susceptible-infection-recovery (SIR) [13]. For example, the SI model assumes that all mobile devices are susceptible to malware attacks (such as malware infections) and get infected after communicating with an infected mobile device. The SIS model considers that a mobile device gets infected, cannot develop an infection immunization, and so possibly becomes susceptible to new malware attacks/infections. However, for the SIR model, the R compartment stands for the recovered compartment that represents the recovered infected mobile devices which are no more susceptible to the malware attack. In reality, IoT devices are continuously vulnerable to new malware attacks, so they rarely develop system immunity. The differential equations are used for the dynamical mathematical systems. For example, the following system of differential equations is used for the SIS epidemic model [9]:

$$\frac{dS(t)}{dt} = \frac{\beta S(t)I(t)}{N}$$

$$\frac{dI(t)}{dt} = \frac{\alpha S(t)I(t)}{N}$$

For this model, N represents the total number of devices, where N equals the summation of devices at the infected and susceptible compartments, $N = I + S$. The parameter β denotes the rate at which the new infected devices occur after the communication between susceptible and infected compartments. On the other hand, the transition rate α is used that reflects the fraction of devices that return to the susceptibility condition after getting infected.

9 Necessities of Big Data Analytics

There is an immense amount of knowledge generated from the whole value chain, comprising the supply of raw materials, manufacturing, delivery of goods, logistics, and consumer aid. It is crucial to comprehensively evaluate 'big data' to extract essential and insightful details [14].

9.1 Improving Processes and Production in Factories

Predictive analytics of supply and consumer data on demand will help increase machinery utilization and improve plant operations.

9.2 Reducing Downtime on Computers

The standard sensors deployed along the entire product line will gather various statistics representing the position of machinery.

9.3 Product Quality Management

On one hand, the study will utilize business demand and consumer demand to enhance the package architecture, ensuring it reflects product changes effectively.

9.4 Enhancing Productivity in the Supply Chain

The propagation during procurement, processing, and transshipment, numerous sensors, RFID, and tags create a vast supply chain predict data that would be used to evaluate supply risk, delivery time, optimum strategic route preparation, etc.

10 Benefits of Big Data Analytics for MIoT

An autonomous vehicle could monitor the road conditions on-ramp to its destination to ensure it complies with traffic regulations. The Internet of Things can also provide services to humans like a house that would adjust its energy budget to house occupants at home or allow the home user to control an appliance remotely through smartphones.

Data analytics helps capture, cleanse, and analyze machine data to reveal insights that can improve performance and improve customer satisfaction. Analytics can also be used to target marketing communications. With analytics, information is gathered on customers, employees, environment, competitors, and general conditions. They use this knowledge to customize communications to individual customer needs. Companies sometimes want to know, for example, which customers are loyal to them and to what extent (Shah et al. 2020). If analytics can measure what motivates customers to return, it can create different promotional messages and offer new ones to those who have been disloyal. It is just one example of the power of analytics. An early example of an analytics project in action was the launch of a campaign to identify the most suitable places to celebrate Princess Diana's birth to be included on the film's official Blu-ray (Shah et al. 2020). This ability to explore and understand the manufacturing processes and supply chain relationships makes it feasible to predict and respond to manufacturing problems before they occur. It provides an overview of how manufacturing processes can be organized to optimize product performance and meet product needs (Shah et al. 2020). It concludes with some considerations for designing, implementing, and evaluating manufacturing processes (Shah et al. 2020). The fact that critical business data is stored beyond the organizational firewall increases severe preoccupations. Four of the most prominent issues are discussed below.

11 Cost Reduction

Hadoop and cloud-based analytics help in cost reduction [12]. Since comparative analysis between big data technologies and conventional environments (data warehouses and marts) is impossible due to functionality discrepancies, comparison alone can imply concessions. Companies, for example, use Hadoop clusters for this specific reason and transfer data to organization warehouses as deemed necessary for analytical output applications.

12 Fraud Detection

Companies nowadays use the cloud to deploy big data analytics [12]. Companies can process real-time data and generate analytics due to scalable and flexible hardware available for a reasonable cost. Thus generated business intelligence helps in fraud detection. For example, it allows insurance companies for fault detection.

13 Product Quality

Product quality is fundamental to the success of IoT systems and their application. An organization's quality is determined in multiple ways. It is a quality attribute if an organization does not have the capability of product quality. However, it can provide an essential insight into a business's overall business performance and customers' needs. The Internet of Things is not a new area of business concern, as demonstrated by the proliferation of sensor networks with their ability to gather real-time data and make predictions on the use and consumption of specific product components. Product quality management is a way to monitor, manage, and optimize the products produced by an organization to ensure quality at the individual product level.

14 Supply Chain Optimization

Supply chain optimization is a technology that helps us take big data as our own and use this as a source of new insights. Imagine the power of using analytics and machine learning to identify new opportunities in the supply chain. Supply chain optimization in big data analytics for manufacturing Internet of Things is a business intelligence approach for maximizing the impact of analytics across multiple supply chains. They use social, physical, and data-driven capabilities. By examining relationships and knowledge in various supply chains, data-driven activities can identify and optimize optimal actions that benefit customers and ensure ongoing value creation for the business (Jiang 2019).

15 Demand Forecasting

Demand forecasting is trying to forecast the future or the next event that may affect the business. Demand forecasting is what allows companies to predict their future production levels at specific points in time. This data is beneficial in forecasting which parts of the supply chain are likely to be the most important for the next few years. Demand forecasting is a process of capturing information about the consumers' consumption patterns and preferences and the consumers' use of the service or services. This data is then fed back into other organization areas, such as service and product innovation, to improve product selection, improve distribution and quality control, improve service quality, and reduce fraud and abuse (Jiang 2019).

In addition to data mining, cloud infrastructure will also keep organizations competitive by offering several advantages, such as cost efficiency, virtualized resources, on-demand operation, rapid elasticity, and ease of administration. Despite these advantages, there are some difficulties and disadvantages, particularly in terms of privacy and protection. A company needs to thoroughly appreciate the scope of what is involved before engaging in cloud-based big data technology. It may be lucrative for a company to engage in cloud analytics, but careful preparation is necessary to ensure that all stages of analytical dimensions are covered.

16 Helps Increase the Businesses' ROI

Companies prefer big data cloud storage with the increased demand for data storage, which ultimately lowers the implementation cost (Lavalle et al. 2020). While some companies decide to use ample data storage only in data centers, others move into cloud computing for some time. A recent survey showed that 85% of companies using cloud storage have decided to open their data to the competition. Cloud computing allows a company to rent computing power from a centralized computing center for less than a percent investment. It provides the ability to store and process large amounts of data for the end user (Lavalle et al. 2020).

17 Advantages to Manufacturing Companies

If manufacturing companies install IoT sensors within their equipment, they can collect critical operational data on the machines (Lavalle et al. 2020). Finally, there are two different types of technical training courses, and there are several reasons for this. Most firms have established systems that provide technical support and advice and involve developing new products, marketing activities, and business processes. Therefore, there is a need for an in-depth look at technical problems (Lavalle et al. 2020).

Data Reliability Although big data is never 100% accurate, it is essential to ensure that the sensors work correctly before analyzing it (Shah et al. 2020). The data's quality that needs to be analyzed is reliable and unaffected by various factors like temperature and humidity. Analyzing data for validity when information is collected and used for something important like improving service is essential to consider its facts and consistency (Shah et al. 2020). When a user logs in or selects a website, the data is stored, processed, and analyzed in real time to enable service.

18 Analysis Depth

To start with, it can quickly learn the data structure, making it more accurate and reliable (Shah et al. 2020). In other words, it learns very quickly, and with this, it can get better at analyzing complex data streams. In addition to automatically teaching data structures, deep learning may have the capability to build complex abstractions, which are not very different from those made with other machine learning techniques. For example, one could build an abstract model such as a prediction tree, in which the nodes in a tree correspond to class labels or attributes that have been predicted. Another example is an automatic credit score-based model based on credit rating used to make loan decisions (Shah et al. 2020).

19 Security

Connected things in various sectors make life easier, but, at the same time, there are significant concerns about data security. Cybercriminals can access data centers and devices; connect to traffic systems, power plants, and factories; and steal users' data. IoT big data is a new phenomenon for security specialists, and the lack of relevant experience increases security risks that should be taken into account. In 2014, an attack against the AWS cloud service revealed vulnerabilities in the storage management infrastructure, and the developers and operations teams needed to take charge of that. The two components of the cloud storage and computing are two widely used technologies (Shah et al. 2020). The security needs for security to balance their security benefits and capabilities are often highly critical to the safety of a company's infrastructure (Shah et al. 2020).

20 Improving Factory Operations and Production

The predictive analysis of manufacturing data predicts customer demand and then keeps machinery and raw materials ready for manufacturing [3]. For example, manufacturing of down coats can be made ready, forecasting a cold wave. Due to this, factory operations and machinery are best utilized.

21 Reducing Machine Downtime

Sensors deployed all over the assembly line collect data related to the machines' status [3]. By analyzing the root cause of a machine failure, steps can be taken to avoid failures to reduce machine downtime. It also helps to detect if any machines are overloaded to balance the load among other machines.

22 Improving Product Quality

By analyzing manufacturing data, it is possible to identify defective goods and eliminate the root cause to improve product quality [3]. Further, by analyzing marketing data and customer requirements, the product design can be enhanced.

23 Enhancing Supply Chain Efficiency

The inventory supply can be arranged based on analyzed supplier data [3]. A lot of data gets generated from an increasing number of sensors, RFID, and tags, which on analysis can help reduce the holding time of supplies and meet the customer demand.

24 Monitoring Manufacturing Process

IoT-based sensors combined with big data processing proved to be sufficient to monitor and detect faults in the manufacturing process. The data generated by IoT devices is enormous and without a big data processing capability, it is hard to analyze and retrieve useful information. Fault detection is essential in the manufacturing process to ensure that everything is running fine. By applying DBSCAN to the collected data, outliers can be removed. The random forest method is further used to predict the faults in the manufacturing process [4].

25 Reduction in Energy Consumption and Energy Costs

Data collected through IoT sensors in the energy-intensive manufacturing process is further analyzed by big data technologies to identify wastage in energy (Zhang et al. n.d.). Thus, identified waste can be eliminated, which also reduces energy consumption and energy costs.

26 Reduction of Scrap Rate

Every manufactured product goes through a series of tests to determine whether the product meets the desired standard [6]. Suppose it does not, then the product is scrapped. By analyzing the data generated during the manufacturing process, it is possible to reduce the scrap rate.

27 Challenges of Big Data Analytics for MIoT

Big data analytics is the most popular and frequently used business analytics approach, especially in the big data business arena. There are many reasons for introducing big data analytics to the analysis of manufacturing integrated manufacturing systems. These include the increased demand for real-time information in manufacturing, the increased awareness of risks associated with those products, and the need to incorporate this information into decision-making. The increased demand for accurate inventory control, coupled with heightened focus on risk management, and the imperative to continually improve quality, have become paramount concerns in today's business landscape. The risk managers must also be concerned that their ability to achieve the objectives requires that they be part of an organization that can respond to their requests. A big data analytics for manufacturing Internet of Things is an IoT device with its operating system, operating system–related firmware, and application program. Its own battery life is combined to produce a more efficient and efficient IoT device for manufacturing (Majeed 2019).

The big data–enabled manufacturing IoT device aims to perform and analyze business analytics to understand the market better and develop more efficient, competitive, and competitively designed products. The new generation of big data-enabled products, sensors, and services will enable businesses to provide better services and to do things themselves that are more reliable, more cost-effective, and more transparent (Majeed 2019). The industry is driven by the need to have a scalable approach and use large amounts of data to understand product behavior and market demand. An example of this is using artificial intelligence to produce a robot in less than an hour with just a few instructions. It is equipped with sensors that can sense traffic conditions to move along a path autonomously or manually [10].

28 Easing Security Concerns

IoT is more secure than the regular Internet or LAN connection; it is not precisely the bulletproof shell some users expected when deploying the product. IoT solutions such as this tend to run on commodity hardware and often lack the sophistication to support cloud-based solutions (Mishra and Chakraborty 2020). This is also true of hardware and firmware-based IoT devices. There are some excellent reasons why IoT devices are susceptible to vulnerabilities: Wiring to the physical network requires regular maintenance and uncontrollable error conditions. No proper security measures are in place that protect the devices from attacks. The IoT relies on both external sources of power and power consumption. It means that the various technologies' risks have to be considered when selecting the appropriate products.

29 Overcoming Connectivity Issues

A centralized, server-client model is used by IoT devices to provide connectivity to the various servers, workstations, and systems in their current form. The IoT is designed to be easily controlled and managed by end users (Mishra and Chakraborty 2020). As such, it is vulnerable to attack vectors that can be used to perform network and application compromises, viruses, denial-of-service attacks, man-in-the-middle attacks, and even man-in-the-browser attacks (Mishra and Chakraborty 2020). The general purpose of the IoT is to enable a seamless integration of cloud and on-premise applications. The primary benefit of using the IoT is the ability to create and maintain security-related data, such as sensor data, firewalls, and alarms. As the cloud computing concept matures, the focus shifts to the deployment of technology-based services that enable the user to deploy, control, and manage a network of compute-capable computing devices (Mishra and Chakraborty 2020).

30 Data Acquisition

Data acquisition deals with collecting data from different sources and representing and transmitting it [3].

30.1 *Data Representation and Transmission*

As data comes in heterogeneous forms, representing the vast data in a standard format is not easy [3]. Similarly, transferring enormous volumes of data to a storage system is also challenging as it requires high bandwidth, and industrial wireless sensor networks have to be highly efficient.

31 Data Preprocessing and Storage

The collected data should be integrated and cleaned to perform analytics [3]. Below are some of the challenges encountered in this phase.

31.1 *Data Integration*

Data in multiple formats from multiple sources is challenging to integrate. The data can be in many forms: structured, unstructured, and semi-structured data [3].

31.2 *Redundancy Reduction*

Data collected from different devices might have temporal and spatial redundancy that might affect the analytics [3]. Therefore, eliminating data redundancy is not easy.

31.3 *Data Cleaning and Data Compression*

Due to defects in the machinery or sensor errors, the collected data might have noise and errors [3]. Therefore, it is crucial to clean the data before performing the analytics to be accurate. But the volume of data generated through manufacturing IoT devices is vast. Therefore, efficient schemes have to be designed to compress the data and then clean the data for analytics purposes. Further discussed are the challenges in data storage.

31.4 Reliability and Persistence of Data Storage

One of the big data analytics requirements is the reliability and persistence of the data [3]. It is challenging to meet the requirement because storing tremendous amounts of data also has a cost.

31.5 Scalability

As data is generated by various machines, sensors are in huge numbers, and the storage to save the data is continually expanding (Yu et al. 2020). In addition to the storage, the tools and techniques applied to the big data should also be scalable. For example, machine learning techniques and open-source tools should also be scalable.

31.6 Efficiency

There is a prodigious number of queries that access the data concurrently [3]. Therefore, data storage should be efficient and reliable, and persistent in supporting the analytics.

32 Data Analytics

After collecting data, preprocessing, and storing, the last phase is to generate data analytics, which further presents the below challenges.

32.1 Data Temporal and Spatial Correlation

The data from conventional data warehouses are different from the MIoT data. MIoT data is spatially and temporally correlated [3]. Extracting valuable information from such data is challenging.

32.2 *Efficient Data Mining Schemes*

The conventional data mining schemes may not apply to MIoT data because of its volume and nature [3]. Efficient data mining schemes have to be designed to manage the tremendous volume and uncertainty of MIoT data.

32.3 *Privacy and Security*

Data privacy and security should be taken care of while generating analytics. The conventional schemes cannot be applied to MIoT data due to its volume, heterogeneous structure, and spatio-temporal correlations [16].

33 Conclusion

With an increased amount of data gathering on the Internet, analysts will see many applications for these tools, such as text mining, statistical models, and machine learning. It will be exciting to see where all this information is headed and what new applications exploit the potential. Some data sources contain real devices like smartphones and tablets and IoT devices that store smart devices, such as smart TVs and intelligent firewalls. The data sources can be in the form of either textual or video data. The main challenge with multimedia data is that unlike text data stored on a mobile device, which will need to be transferred between devices, there is no data stored on a desktop computer. So, we are going to need a mobile client with a proper application. Mobile devices play an essential role when making transactions for mobile clients. If you are making a phone call or going to buy or rent a car, you will likely need to make these transactions from your laptop. You cannot do it from your mobile device.

Many analysts and big data professionals are now using predictive analytics as part of their regular job functions, just as banking. Therefore, the benefits of data analytics will, once again, be felt throughout organizations, not just the ones doing big data. We can also apply these principles of predictive analytics and big data management to our healthcare business, which will bring significant value and benefits for our patients and the healthcare provider as a whole [15]. Data security is one of the most fundamental technologies and principles of big data, as it is directly impacted by the fact that we are accessing large amounts of unstructured data every second. Data security can also be related to information security and privacy. A secure user's data is an asset of the organization, and it needs to be protected with appropriate mechanisms.

This chapter discussed various benefits that industry owners could experience by applying big data analytics to MIoT data. It also presents the classification of multiple challenges that are encountered while generating the analytics. Although there are challenges involved, the benefits add a lot of value to the industrial process.

References

1. Bahwaireth, K., Tawalbeh, L.A., Benkhelifa, E., Jararweh, Y., Tawalbeh, M.A.: Experimental comparison of simulation tools for efficient cloud and mobile cloud computing applications. EURASIP J. Inf. Secur. **2016**(1), 1–14 (2016)
2. Cochran, D.S., Kinard, D., Bi, Z.: Manufacturing system design meets big data analytics for continuous improvement. Procedia CIRP. **50**, 647–652 (2016). https://doi.org/10.1016/j.procir.2016.05.004
3. Dai, H.-N., Wang, H., Xu, G., Wan, J., Imran, M.: Big data analytics for manufacturing internet of things: opportunities, challenges and enabling technologies. Enterprise Information Systems. **14**(9–10), 1279–1303 (2020). https://doi.org/10.1080/17517575.2019.1633689
4. Hong-Ning Dai, H.W.: Big Data Analytics for Manufacturing Internet of Things: Opportunities, Challenges and Enabling Technologies, pp. 2–8 (2019)
5. Krumeich, J., Jacobi, S., Werth, D., Loos, P.: Big data analytics for predictive manufacturing control – a case study from process industry. In: 2014 IEEE International Congress on Big Data. https://doi.org/10.1109/bigdata.congress.2014.83 (2014)
6. Lade, P., Ghosh, R., Llc, R.B., Srinivasan, S.: Manufacturing analytics and industrial internet of things. IEEE Intell. Syst. **6** (2017)
7. Liu, Z.Y., Guo, Y.B.: Big data analysis for smart manufacturing. In: Big Data and Computational Intelligence in Networking, pp. 499–514 (2017). https://doi.org/10.1201/9781315155678-24
8. Mahboubi, A., Camtepe, S., Ansari, K.: Stochastic modeling of IoT botnet spread: a short survey on mobile malware spread modeling. IEEE Access. **8**, 228818–228830 (2020). https://doi.org/10.1109/ACCESS.2020.3044277
9. Martin, J.C., Burge, L.L., Gill, J.I., Washington, A.N., Alfred, M.: Modelling the spread of mobile malware. Int. J. Comput. Aided Eng. Technol. **2**(1), 3–14 (2010)
10. Muheidat, F., Tawalbeh, L.A.: Artificial intelligence and blockchain for cybersecurity applications. In: Artificial Intelligence and Blockchain for Future Cybersecurity Applications, pp. 3–29. Springer, Cham (2021)
11. O'Donovan, P., Leahy, K., Bruton, K., O'Sullivan, D.T.: Big data in manufacturing: a systematic mapping study. J. Big Data. **2**(1) (2015). https://doi.org/10.1186/s40537-015-0028-x
12. Prasad, B.M.: Challenges and benefits of deploying big data analytics in the cloud for business intelligence. Procedia Comput. Sci., 6–10 (2017)
13. Rey, M.D.: Mathematical modeling of the propagation of malware: a review. Secur. Commun. Netw. **8**(15), 2561–2579 (2015)
14. Siegel, E.: Predictive Analytics: the Power to Predict Who Will Click, Buy, Lie, or Die. Wiley (2013)
15. Tadaka, S.M.: Applications of blockchain in healthcare, industry 4, and cyber-physical systems. In 2020 7th International Conference on Internet of Things: Systems, Management and Security (IOTSMS), pp. 1–8. IEEE (2020, December)
16. Tawalbeh, L.A.A., Mohammad, A., Gutub, A.A.A.: Efficient FPGA implementation of a programmable architecture for GF (p) elliptic curve crypto computations. J. Signal Process. Syst. **59**(3), 233–244 (2010)
17. Viktor Mayer-Schönberger, K.C.: Big Data: a Revolution that Will Transform how we Live, Work, and Think. AN Eamon Dolan Book, Boston/New York (2015, March)

Artificial Intelligence-Based Secure Edge Computing Systems for IoTDs and Smart Cities: A Survey

Muhammad Asim, Chen Junhong, Liu Wenyin, and Ahmed A. Abd El-Latif

1 Introduction

With the advancement of Internet of Things devices (IoTDs) and wireless communication systems [1], a huge number of physical entities are being connected by IoTDs to provide services to users [2–4]. It is the Internet that allows and connects objects to communicate and exchange data, such as sensors, smart meters, Internet-connected vehicles, smartphones, and other Internet-connected objects [5]. The interconnection and communication among IoTDs lead to the advancement of appli-

M. Asim (✉)
School of Computer Science and Technology, Guangdong University of Technology, Guang, PR China

EIAS Data Science Lab, College of Computer and Information Sciences, Prince Sultan University, Riyadh, Saudi Arabia
e-mail: asimpk@gdut.edu.cn; masim@psu.edu.sa

C. Junhong
School of Computer Science and Technology, Guangdong University of Technology, Guangzhou, PR China

Expertise Centre for Digital Media, Hasselt University, Hasselt, Belgium

L. Wenyin
School of Computer Science and Technology, Guangdong University of Technology, Guangzhou, PR China

A. A. Abd El-Latif
Center of Excellence in Quantum and Intelligent Computing, Prince Sultan University, Riyadh, Saudi Arabia

Department of Mathematics and Computer Science, Faculty of Science, Menoufia University, Shebin El-Koom, Egypt
e-mail: aabdellatif@psu.edu.sa

A. A. Abd El-Latif et al. (eds.), *Secure Edge and Fog Computing Enabled AI for IoT and Smart Cities*, EAI/Springer Innovations in Communication and Computing,
https://doi.org/10.1007/978-3-031-51097-7_13

cations, such as tracking of products, Internet banking [6], environmental and health monitoring, and the automation advancement in our daily life [7]. Smart home is one of the applications of IoTDs [8, 9]. In smart homes, residents can automatically open their garages on arrival to home, switch on their air conditions, switch on/off lights, and control other appliances [10–12]. IoTDs have many other applications in various domains, such as smart grid, intelligent transportation, e-healthcare [13–15], industrial automation, and disaster management. The advancement of IoTDs applications open doors for the innovations that connect human with IoTDs and provide opportunities, infrastructures, and enhance the quality of services for the betterment of daily life. Some of these IoTDs' applications are latency-sensitive and may have private data that need to be processed and stored locally, while some contain a huge amount of data, that may put a heavy burden on networks [16]. Such data may consist of multimedia information, such as photos, voices, videos, etc., that may lead to network congestion and put a heavy load on IoTDs. However, due to the limitation of IoTDs like limited computation resources and battery, such applications and data cannot be processed and stored on them.

To address the above mention issues of IoTDs, cloud computing (CC) has been introduced. CC can provide on-demand services to IoTDs. It provides centralized storage, computing, and resources in the remote clouds [17, 18]. It is a centralized architecture that allows pertinent, permeating, and on-demand access to shared pools of configurable resources [19]. In CC, IoTDs offload their tasks to the remote cloud, which processes their tasks and returns results to them. Although CC furnishes unlimited computation resources with high accessibility for IoTDs, however, it may fail to fulfill the requirements of latency-sensitive applications which require high reliability. Furthermore, due to the physical distance between remote/central cloud and IoTDs, it is more vulnerable to security threats [20].

To overcome the above-mentioned challenges in CC, edge computing (EC) has been proposed, which provides resources at the user end in a decentralized way and responds to IoTDs much faster than CC [21]. It can enable analytic and knowledge generation to occur at the source of the data close to the user end [22]. It leverages resources which may not be continuously connected to the network such as sensors, tablets, laptops, and smartphones as can be seen in Fig. 1. Figure 1 presents the architecture of CC and EC. Being close to the user, EC responds to users' requests faster and more securely than CC [23]. Although some existing solutions in CC may be shifted to handle some security issues in EC. However, it still inherits some specific security challenges because of the distinctive features of EC servers such as location awareness, mobility support, etc. EC provides more secure services to IoTDs compared to CC as it can store/process data locally. EC servers may work as proxies for providing secure services to IoTDs, if they lack adequate resources for processing their tasks [24].

However, as an extraordinary extension of CC, it still inherits a number of issues, especially security and privacy issues [25, 26]. EC has been deployed via different edge servers which are not fully trusted and thus IoTDs are vulnerable to be attacked, as a result, EC servers face different kinds of security and privacy threats [27, 28]. Unfortunately, these security issues in EC have not been systematically identified.

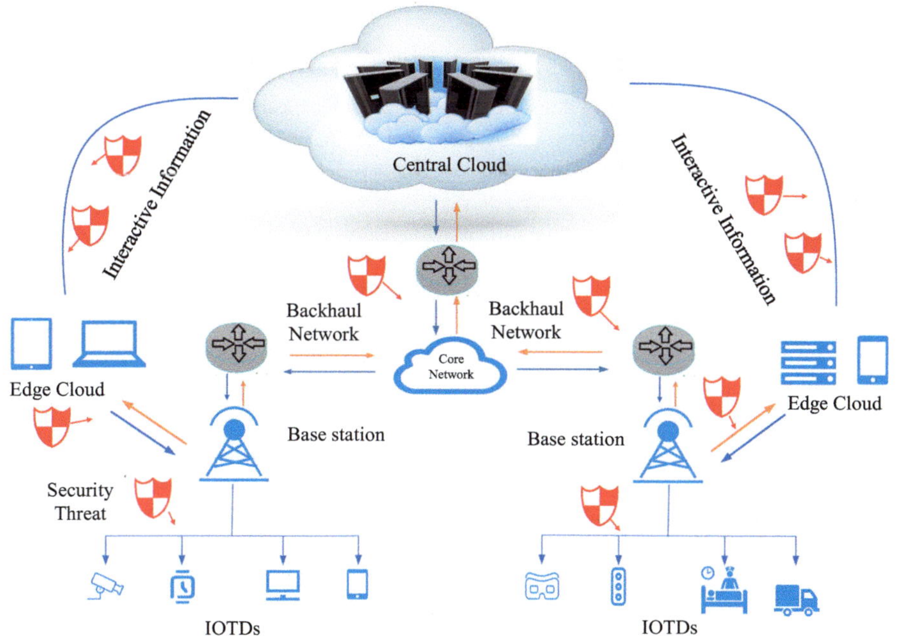

Fig. 1 Architecture of edge computing systems [21]

AI displays excellent potentials in addressing security issues of EC, which attracts great attention recently from industry and academia. It has evolved due to large-scale development and knowledge generation through sensing systems, IoTDs, social media, and web applications. Such data can be handled by using various learning techniques, which often depend on centralized training models using central clouds provided by companies such as Google, Amazon, Oracle, etc. Fuzzy logic-based algorithms (FLAs) and learning-based algorithms (LBAs) are the two most important systems that are often applied to handle security issues in EC.

The main purpose of our chapter is to review AI techniques used to address security issues in EC. Several works have been devoted to AI and EC and their related topics, for example, Elgendy and Yadav [29] presented the taxonomy of computational offloading approaches in mobile edge-cloud computing. Singh et al. [30] presented a comprehensive survey that described the security and privacy issues in various layers of the EC architecture. Asim et al. [21] discussed the use of computational intelligence (CI) techniques for addressing issues like task offloading, resource allocation, etc., in the cloud and edge computing. Lansky et al. [31] focused on the deep learning (DL)-based intrusion detection schemes and provided a systematic survey of these schemes in network communications. Zargar et al. [32] presented a survey of defense mechanisms against distributed denial-of-service flooding attacks. Masdari and Ahmadzadeh [33] focused on the comprehensive analysis of the authentication methods in wireless body area networks. Sun et al.

Table 1 Survey papers in EC

Ref.	Year	Published in	Focus
[29]	2021	Springer International Publishing	MEC, CC
[30]	2021	IEEE Internet of things Journal	Machine learning, fog EC
[21]	2020	IEEE Trans. on emerging topics in CI	CI, CC, EC
[31]	2021	IEEE Access	DL-based intrusion detection systems
[32]	2013	IEEE Comm. Surveys & Tutorial	Distributed denial-of-service
[33]	2016	Security and Comm. Networks	Wireless body area networks
[34]	2019	IEEE Access	E-Healthcare system
[35]	2020	Jour. of Sys. Arch.	Wireless ad-hoc networks
[36]	2020	App. Soft Computing	FLAs based intrusion detection systems
[37]	2019	IEEE Comm. Surveys & Tutorial	DL, Mobile and Wireless networking
[38]	2020	IEEE Comm. Surveys & Tutorial	EC, DL
[39]	2015	Springer International Publishing	Security and privacy in Fog
[40]	2021	Cluster Computing	Security anomaly in SDN
[41]	2018	ACM Computing Surveys	Anomaly Detection, DL
[42–48]			MEC
Our work	2021		AI, EC, and AI for security issues in EC

[34] reviewed the security and privacy challenges, requirements, threats, and future research directions for the Internet of medical things enabled healthcare systems. Khan et al. [35] presented a survey on intrusion detection and prevention in wireless ad-hoc networks. Masdari and Khezri [36] presented a comprehensive survey of the fuzzy misuse detection schemes designed using various machine learning and data mining techniques to deal with different kinds of intrusion detection systems. Zhang et al. [37] reviewed DL in mobile and wireless networking. Wang et al. [38] surveyed convergence of EC and DL. Yi et al. [39] surveyed security and privacy issues in fog computing. Jafarian et al. [40] presented a survey and classification of the security anomaly detection mechanisms in software-defined networks (SDN). Pang et al. [41] reviewed DL for anomaly detection. Other surveys specifically targeting the EC and its related issues and paradigms are [42–48]. Table 1 presents all the above discussed surveys and reviews at a glance along with the and fields they are focused.

However, none of the above studies focused on the use of AI techniques for handling security issues n EC. The motivation behind this chapter is to provide a detailed survey of AI algorithms and techniques used to address security issues in EC. In addition, it also presents some paradigms of AI and EC in detail. This chapter focuses on summarizing the work on using AI techniques to address and resolve security-related problems in EC, rather than a comprehensive survey of AI or EC.

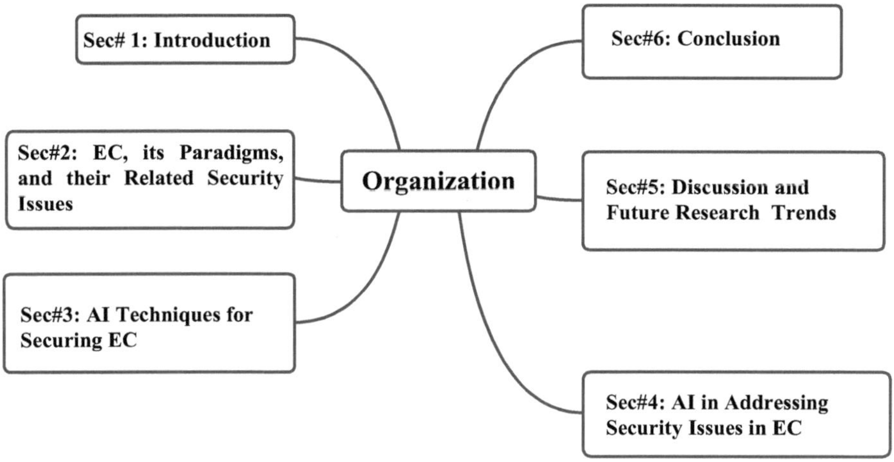

Fig. 2 Chapter organization

1.1 Chapter Organization

The remaining sections of this paper are organized as follows. In Sect. 2, we present an overview of the concepts related to EC and their related security issues. Section 3 introduces AI techniques applied for handling security challenges in EC. In Sect. 4, we elaborate on a brief review of the existing work dedicated to the application of AI techniques in EC. Finally, in Sect. 5, we offer possible future challenges and research trends for handling security issues in EC, followed by the conclusion of the chapter in Sect. 6. For more explicit clarity, the pictorial presentation of the organization of the chapter is described in Fig. 2.

2 Edge Computing

EC was first proposed by Vaquero et al. [49]. It pushes the frontier of computing applications, data, and services from the central cloud to the network edge (i.e., users' end). This approach needs leveraging resources that may not be linked to the network. In addition, it facilitates downstream data at the user's end on behalf of central clouds and upstream data on behalf of IoTDs by providing computing and network resources along the path between data sources and cloud data centers [50]. EC provides real-time distributed services with low latency, high mobility, and location awareness. However, it furnishes less capacity of resources, compared to CC. EC is a general term for many computing paradigms, for example, fog computing, mobile EC (MEC), and cloudlet. The main paradigms of EC are given below.

- **Cloudlet:** It was first proposed by Satyanarayanan et al. [51, 52]. It has been introduced to provide resources in a widely distributed and decentralized manner near to IoTDs. It is known as the "data center in a box." It can quickly execute and process latency-sensitive tasks for the IoTDs without referring to remote clouds to meet the requirements of latency-sensitive applications [53]. Although, it has several advantages of being near to IoTDs and can be exploited by IoTDs even if they do not have an Internet connection but still its potentials are very limited compared to the central cloud. Therefore, the computing power of cloudlet may be limited in providing specific services.
- **Fog computing:** It was first discussed by Cisco System in 2012 [54] and is referred to as a computing paradigm furnishing storage, computing, and resources close to IoTDs. It can reduce the data amount sent to the latter for processing and storage by extending the services of central clouds to users' end [55]. Moreover, it can also be integrated into the radio system in the mobile networks operators [56]. Furthermore, it can exploit storage functions, calculate, control, communications between IoTDs, depending on the type of users requests [57].
- **Mobile EC:** The latest concept of EC is mobile EC (MEC). This architecture has been introduced in 2014 by the European Telecommunications Standards Institute (ETSI), which provides computation resources through cloud servers close to IoTDs via the radio networks [39]. According to ETSI, this architecture reduces the latency considerably and better locates user position. It improves the performance of the network via providing various services (e.g., software-defined network) to IoTDs [58]. It aims to improve the bandwidth, reduce the latency and jitter adequately, and make available the quality of service (QoS) for mobile applications by integrating services closer to IoTDs.

2.1 Security Issues in EC

Security is a big problem for IoTDs to execute/send their tasks either to/at the central cloud or to/at the nearby edge cloud. A huge number of hackers are trying to hack/steal/interrupt data while IoTDs execute their tasks on the nearby edge cloud. It is a set of control-based technologies and policies designed to adhere to regulatory compliance rules and protect the information, data applications, and infrastructure associated with EC.

As a non-trivial extension of CC, EC is considered to be a more secure architecture than CC. The reasons why EC is more secure than CC are as follows: (1) In EC, the tasks of IoTDs are processed and analyzed on local closest edge clouds, which reduces their dependency on the Internet connections. Locally sharing, processing, and storing data make it challenging for hackers to hack/steal data of IoTDs/users. (2) The exchange of data and information between the IoTDs and the central clouds is not required in the real-time processing of data, therefore, making it difficult for hackers to steal/disrupt the information/data of IoTDs.

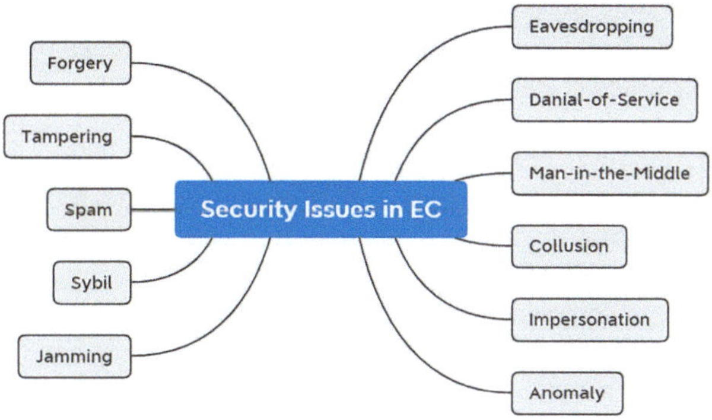

Fig. 3 Security issues in EC

However, EC inherited several security threats like forgery, spam, tampering, jamming, etc. [59]. Some of the security issues in EC are listed in Fig. 3 and are discussed below in detail.

- Forgery: Many malicious attackers and hackers not only falsify their identities but also produce fake/false information to misguide users and server providers. Moreover, such attacks lead to the excessive consumption of network resources, like bandwidth, storage, and energy.
- Tampering: Such types of attackers may maliciously delay, drop, or modify the data transmission of IoTDs to interrupt EC clouds and demote their efficiencies. Some of the tampering behaviors are difficult to be detected, as the condition of wireless channel and mobility of IoTDs may disconnect/delay the transmission of data.
- Spam: It is referred to as the unwanted contents generated and spread by hackers and attackers, for example, redundant information, false transmission, and collection of data from IoTDs. It misleads social communications and friends and even result in privacy leakage.
- Sybil: It is referred to as an attacker that manipulates fake identity or abuse pseudonyms to control/compromise the effectiveness of EC. For example, the attacker might generate incorrect crowdsensing reports, for example, the results of crowdsensing may not be trustworthy. Moreover, the attacker can disrupt authorized IoTD's sensitive information [59].
- Jamming: Such attackers deliberately generate a huge number of bogus messages that may jam/block the communication links between IoTDs and EC clouds. In addition, it can also block the computing resources of EC clouds, as a result, IoTDs are prohibited to process their tasks in a normal way.
- Eavesdropping: Such malicious attackers may listen and interrupt the communication channels between IoTDs and EC clouds, as a result, they may capture the transmitting packets and read the contents of IoTDs and users.
- Denial-of-Service (DoS): The attackers may disrupt the services provided via EC clouds to make them unavailable to its intended IoTDS, by sending superfluous requests to the target EC clouds. Such types of attackers consume extra network

resources to interrupt and prevent the tasks of IoTDs from being processed. An EC cloud is more vulnerable to the DoS attacks than the central cloud due to its limited resources availability [59].

- Man-in-the-Middle: Such malicious attackers stand in between (i.e., in the middle of) two parties (i.e., IoTDs and EC clouds). These attackers can secretly relay/modify the data exchange between two parties, where both of them think that they are communicating directly with one other.
- Collusion: It occurs when two or more parties (i.e., EC clouds or IoTDs) collude together, which may deceive/mislead other legal entities. In EC, two or more parties, such as EC clouds, IoTDs devices, etc., may collude in order to increase their attacks capability.
- Impersonation: Such types of malicious attackers pretend legitimate IoTDs/users to enjoy the services of EC clouds provided to IoTDs. In addition, it can impersonate a legitimate EC cloud to offer phishing services to IoTDs.
- Anomaly: It is referred to as an observation violating standard patterns or deviating significantly from the overall specification of the data [60].

3 AI Techniques for Securing EC

AI is an area of intelligent system, which is referred to a set of computing and intelligent methods that can solve complex real-world problems in various research fields. It describes the capability of machine learning just like humans and the ability to respond to certain behaviors. AI has been attracted significant attention and the importance of researchers around the globe in all fields of science and engineering. Specifically, the role of AI techniques in addressing security issues in communication systems is expanding with the growth and advancement of the Internet and IoTDs. AI techniques play a key role in handling national-security challenges and are used in human well-being-critical systems. In the Internet era and communication system, humans/users face fishing attacks, which are solved by AI through automatic phishing detection. In addition, IoTDs suffer from impersonation and insecure data collections and sharing, which can be handled by adding AI layers in the central cloud or EC cloud. Moreover, AI has been used as a logic-based framework and policy for handling cyberattacks in critical infrastructures.

In this work, we focus on just two categories of AI techniques that are often applied in EC for addressing security issues in EC, i.e., FLAs and LBAs.

3.1 Fuzzy Logic-Based Algorithms

FLAs are a type of AI family, which mainly include two types, namely, fuzzy logic and fuzzy inference system in the context of addressing security issues in EC.

- **Fuzzy Logic (FL)**: It was first proposed by Lotfi Zadeh in 1965 [61]. It follows the process of approximate reasoning rather than the process of the exact approximation. It is based on a fuzzy set which is an extension of the classical set. It is inspired by the natural language and thinking ability of humans which make it different from traditional logic. In the theory of traditional logic, a member has a logic value of either 0 or 1 (i.e., false or true), which means that either the member belongs to a given set or not. Thus, the traditional logic theory has no partial membership and has only a white and black concept with less expected obscurity[62]. Fuzzy modeling can provide the best alternatives for this fuzziness and obscurities as they make the nonlinear or complex systems unfeasible, whereas numerical models cannot handle these ambiguities under such situations.
- **Fuzzy Inference System (FIS)**: FIS model infers given crisp values as inputs selected from an input set, their fuzzy set, a membership function, and a set consisting of inference rules [21, 63]. Mamdani and Takagi-Sugeno models [63, 64] are the most commonly used models. FIS rules are normally settled by expert knowledge or by learning from available data. This kind of system can be constructed as: First, the inputs for the system are fuzzified by using the membership function which is linked to a pre-defined fuzzy set. The fuzzy set is labeled by using linguistics terms. These linguistic terms are used by the inference rules as antecedents and/or consequents. Then, the firing strengths are calculated for all rules, which shows the importance of rules in approaching the conclusion. After that, for all rules, the above-mentioned strengths are aggregated and weighted which produces a fuzzy value. At the final stage, the fuzzified values are then defuzzified one by one via applying a suitable method [21].

3.2 Learning-Based Algorithms

LBAs are inspired by the learning behavior of living organisms. LBAs are used to reveal/uncover the relationship between nominal features set and states/objects [65]. In the context of security issues in EC, LBAs can be categorized into the following three categories.

- **DL**: DL is a subtype of LBAs proposed by Hinton et al. [66]. It is inspired by the thinking ability of humans that is the signal processing mechanisms of the human brain/neurons. It cataracts layers in order to extract features from the input data and eventually can form a decision [67]. The architecture of a DL system consists of several processing layers, each of which can generate a nonlinear response based on input data [68]. The DL architecture has an input layer and an output layer along with one or more hidden layers. In the input layer, the environment state is represented as an appropriate number which can be considered as input to the network. Then, the activation function, such as the Sigmoid function or the Tanh function, is introduced to represent/interrupt the recollection of the results.

The weight of ownership between layers is then updated. Thus, DL architecture can be constructed in this way.
- **Reinforcement Learning (RL)**: RL [69, 70] is a type of LBAs determining optimal policies to dictate which action to take at a certain state to attain the highest reward. For RL, first the problem is formulated/converted into Markov decision process (MDP), where one has a set of states and actions. The objective of the problem is to discover the associations of optimal action state, which is known as the policy function. In RL, agents in a given state select the appropriate action from the action set, receive a reward value, and then enter to a new state to select an another action in the already-updated set of actions.
- **Deep RL (DRL)**: DRL is a type of LBAs that combines principles of DL and RL to create efficient learning algorithms. These are powerful models that implement DL models with RL models to solve complicated problems in various research areas such as robotics, video games, finance, and healthcare [71]. Deep Q-learning (DQL) and deep Q-networking (DQN) are two commonly used DRL algorithms.

4 AI in Addressing Security Issues in EC

In this section, we provide the AI approaches applied to solve the above-mentioned security challenges discussed in Sect. 2.1.

4.1 FLAs for Security and Privacy in EC

The works found using FLAs for handling security issues/challenges in EC are discussed in this section.

A security service chaining based on fuzzy theory was proposed for sustainable MEC in [72]. The proposed model applied a FIS to achieve optimal results for the studied security functions. Ritu and Shushma [73] proposed a trust model via applying FL in CC, which is an important factor in the area of information technology and security. Chenhao and Rajkumar [74] studied and evaluated a trust system via applying FIS for the selection of services in the cloud. The proposed model evaluated the trust by using the user's fuzzy QoS requirements and performances of dynamic services. Balan et al. [75] proposed fuzzy-based intrusion detection systems in mobile ad-hoc networks. The proposed system detected the malicious behavior of the node via an intrusion detection system based on FL as well as identify the type of attacks. Devi et al. [76] investigated a 5G wireless communication network with an incorporated relay. They implemented an intrusion detection system by proposing an adaptive neuro-FIS for detecting an attack on the relay. Mansour et al. [77] proposed an integrated three-tier trust management framework in mobile edge computing using FIS.

4.2 LBAs for Security and Privacy Issues in EC

Recently, LBAs have attained a lot of significance and attention for addressing security issues in EC. The research focused on the use of LBAs for addressing security issues of EC and IoTDs are listed below.

Chen et al. [78] presented a DL-based approach by offloading tasks to nearby edge to detect malicious attacks and security threats in MEC. A cloud-enabled task offloading architecture based on DRL for vehicular EC networks was investigated in [79], where the authors considered the remote cloud as a backup server. The proposed approach provides effective and efficient results. Ahmad et al. [80] studied a DL-based classification model for botnet attack detection. They proposed a model for detecting botnets using DL to identify zero-day botnet attacks in real-time. Wang et al. [81] DL-based classification and anomaly detection of side-channel signals. Hussain et al. [82] applied a DL for the detection of the anomalies in a MEC system, where the computations of the core network are distributed across the cellular infrastructure among different MEC servers to reduce the burden on the core network. Chen et al. [83] proposed DL model for secure MEC in cyber-physical transportation systems. A DL-based model is designed to learn attack features via using unsupervised learning in order to accomplish the active learning process. Hussain et al. [84] proposed a DL-based architecture for anomaly detection in cellular networks. Masood et al. [85] proposed a DL-based detection of sleeping cells in next-generation cellular networks. They employed a DL-based framework that uses minimization of drive tests (MDT) functionality introduced in LTE networks. In their proposed framework, MDT measurements were used to train the DL model to quickly detect anomalies or cell outages in the network. Chi et al. [86] proposed a practical method for privacy-preservation in DL classification tasks based on bipartite topology threat modeling and an interactive adversarial deep network construction in the context of EC. They called this approach as Privacy Partition. Nagisetty et al. [87] presented a framework for detection of malicious activities in IoTDs backbone networks using Keras DL library. They used four different DL models such as multi-layer perceptron, convolutional neural networks, Deep Neural Networks, and Auto-encoder for predicting malicious attacks. Binbin et al. [88] proposed a security and cost-aware computation offloading strategy based on DRL and deep Q-network for mobile users in a MEC environment, aiming to minimize the overall cost under the risk probability constraints. Gupta et al. [89] proposed a machine learning and smart card-based two-factor authentication scheme for preserving anonymity in telecare medical information systems. Binbin et al. [90] proposed an RL-based security-aware workflow scheduling scheme to solve the workflow scheduling problem in a dynamic MEC facing security threats. Sedik et al. [91] presented a system to detect alterations to biometric modalities to discriminate pristine, adulterated, and fake biometrics in 5G-based smart cities. Specifically, they used DL models based on convolutional neural networks and a hybrid model that combines convolutional neural networks with convolutional long-short term memory to compute a three-tier probability that a biometric has been tempered. Xu et al. [92] a multi-agent federated reinforcement learning for secure incentive mechanism in intelligent cyber-physical systems. Mothukuri et al. [93] presented a federated learning-based anomaly detection for IoTD security attacks. Mighan and Kahani [94] proposed a novel scalable intrusion detection system based on DL. Popoola et al. [95] studied stacked recurrent neural network for botnet detection in smart homes and IoTDs.

5 Discussion and Future Research Trends

The works reviewed in this chapter show that researchers have focused on the innovational integration of AI approaches to handle security issues/challenges in EC. However, some challenges and issues are still open to being handled. This section provides a discussion with some future research trends on the applications of AI in EC.

5.1 Discussion

From the literature reviewed in this chapter, AI techniques are very efficient in addressing security issues in EC. As shown in Fig. 4, most of the works (i.e., 75%) reviewed in this chapter are on the use LBAs for addressing security issues in MEC. However, few works were found on the use of FLAs compared with LBAs. Thus, we conclude that LBAs are more suitable than FLAs for handling security issues. The superiority of LBAs are attributed as (1) LBAs can adapt themselves according to IoTDs' requirements automatically; (2) these algorithms can learn and explore new knowledge from large databases and efficiently prevent the system from various attacks; (3) LBAs have strong abilities to handle complicated security issues via learning from the past experiences.

Fig. 4 The distributions of LBAs and FLAs for handling security issues/challenges in EC

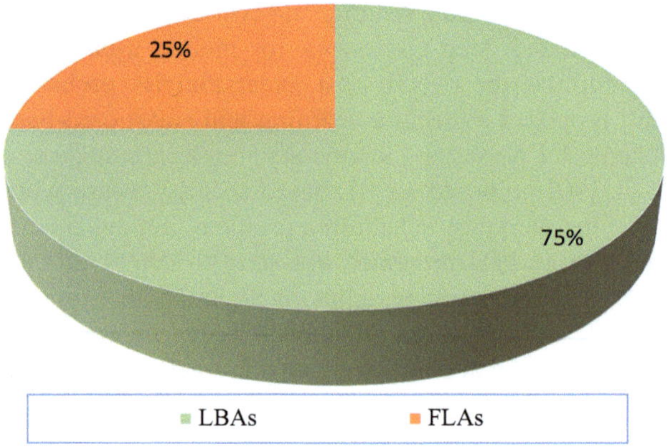

5.2 *Future Research Trends*

5.2.1 From EC Perspective

- Multi-Objective Formulation Issue: Several security issues need to be considered simultaneously in EC in order to generate a trade-off between them. Moreover, in the future, one can take into account the trade-off security issues with some other critical issues such as latency, energy consumption, etc., while satisfying the QoS of IoTDs.
- Hybrid CC and EC: Both CC and EC have their own advantages and disadvantages. For instance, EC has limited resources while CC has huge resources but faces security issues and suffers from long latency. In the future, one can design a secure hybrid system based on the demands of IoTDs.
- Edge Allocation Issue: Because of the distributed nature of EC, it is difficult to assign/allocate edge nodes that provide services across geographic locations and regions. In the future, one can design efficient service discovery protocols for IoTDs so that they can identify service providers based on their needs and requirements.
- Real-Time Joint Optimization: Because of the highly dynamic nature of the service environment for many edge scenarios, it is more vulnerable to malicious attacks. As a result, it requires superior functionality in online edge resource orchestration and configuration to continuously and securely adapt to large-scale edge intelligence tasks. In the future, secure real-time joint optimization of heterogeneous computing, communication, and cache resource allocation is needed to meet dynamic task requirements.
- Decentralized Trust Issues: EC openness leads to decentralized trusts, such as the services furnished via various edge entities that need to be secure and trustworthy. Lightweight and distributed security mechanisms are therefore needed to be designed to ensure EC user authentication and access control, model and data integrity, and mutual platform validation. In addition, it is important to study the new security routing scheme and trust network topology of EC service delivery when considering the coexistence of trusted edge nodes with malicious nodes. IoTDs, on the other hand, generate large amounts of data at the edge of the network that may contain private data of users such as location, activity records, etc. Thus one can design a feasible paradigm in the future to reduce data sharing in order to improve users' privacy.
- Blockchain-Empowered EC: Blockchain can improve the security of the communication and computing resource sharing between IoTDs and edge clouds. In the future, one can design blockchain-empowered EC models to further strengthen the capabilities of EC against security attacks.
- Applications Migration in EC: Migration of EC applications among different edge clouds can lead to security threats. One of the future trends is to design efficient AI techniques and models to handle the applications migration challenges.

5.2.2 From AI Algorithms Perspective

- Multi-Objective Problem: multiple objectives should be considered simultaneously in EC in order to satisfy the QoS requirements of IoTDs. Therefore, in the future, researchers need to develop multi-objectives AI techniques to address many objectives (i.e., many security issues) in EC at the same time.
- Constraints Handling in AI: Most of the AI-based algorithms may not handle constrained problems efficiently, therefore, in the future, one may combine some constraints handling techniques with AI algorithms to address security-constrained optimization problems in EC.
- AI for Data Sharing Security: Sharing data directly across multiple edge clouds can lead to privacy breaches risk. Federated learning has the capabilities to handle data sharing security risks and challenges. Therefore, it can be incorporated to protect and reduce the breaching of data during data sharing via the efficient training of the distributed data. In such models, the original dataset is kept remain in the same nodes it is generated on, while the parameters of the AI model are shared between various nodes.
- Intelligent EC Models: Instead of taking advantage of the already established resource-intensive AI model in EC, one can design some novel resource-aware EC models based on AI. One of the future trends may be to design some novel resource-aware edge AI edge models.
- Hybrid AI: Hybrid AI technology shows great influence in solving problems in various fields, but few papers study hybrid algorithms for solving EC problems. The trend in the future may be to design hybrid algorithms by combining different AI technologies. For instance, one can hybridize FLAs with LBAs to efficiently handle complicated security issues in EC. Similarly, LBAs such as DL, RL, and DRL can be combined with federated learning to reduce data sharing risk in the EC.
- AI-based Collaboration between CC and EC: Several latency-sensitive requests may be received simultaneously in an edge cloud that may not be handled by edge cloud alone. Thus, fast and efficient AI algorithms (e.g., LBAs and FLAs) with some prediction approaches should be designed for collaboration among several edge/remote clouds.
- AI-based Migration of Applications in EC: Sharing and migrating applications between edge clouds can be a challenging task leading to security threats and service delay. In the future, we need some AI-based models for applications migration among edge clouds.

6 Conclusions

In this chapter, we have surveyed applications of AI techniques relevant to critical security issues appertaining to EC and its related paradigms. We presented concepts of EC along with some related paradigms and their security issues, then focus on enlightening two categories of AI techniques often found in EC for handling security issues. Additionally, diverse designed models and approaches based on AI algorithms have been surveyed in the context of EC. The most commonly used and found AI techniques in EC for handling security issues are LBAs and FLAs. However, they are still not fully utilized in EC, therefore in the future, they can become more visible for security issues in EC. In addition, we expect/foresee a huge number of research work on the application of AI techniques in handling diverse security issues in EC.

References

1. Asim, M., Mashwani, W.K., Belhaouari, S.B., Hassan, S.: A novel genetic trajectory planning algorithm with variable population size for multi-uav-assisted mobile edge computing system. IEEE Access **9**, 125569–125579 (2021)
2. Al-Fuqaha, A., Guizani, M., Mohammadi, M., Aledhari, M., Ayyash, M.: Internet of things: a survey on enabling technologies, protocols, and applications. IEEE Commun. Surv. Tutor. **17**(4), 2347–2376 (2015)
3. Khan, A.A., Kumar, V., Ahmad, M., Gupta, B.B., Ahmad, M., Abd El-Latif, A.A.: A secure and efficient key agreement framework for critical energy infrastructure using mobile device. Telecommun. Syst. **7**, e700 (2021)
4. Abd El-Latif, A., Abd-El-Atty, B., Venegas-Andraca, S., Mazurczyk, W.: Efficient quantum-based security protocols for information sharing and data protection in 5g networks. Future Gener. Comput. Syst. **100**, 05 (2019)
5. Gubbi, J., Buyya, R., Marusic, S., Palaniswami, M.: Internet of things (IoT): a vision, architectural elements, and future directions. Future Gener. Comput. Syst. **29**(7), 1645–1660 2013. Including Special sections: Cyber-enabled Distributed Computing for Ubiquitous Cloud and Network Services & Cloud Computing and Scientific Applications ? Big Data, Scalable Analytics, and Beyond
6. Gad, R., Abd El-Latif, A.A., Elseuofi, S., Ibrahim, H.M., Elmezain, M., Said, W.: IoT security based on iris verification using multi-algorithm feature level fusion scheme. In: 2019 2nd International Conference on Computer Applications & Information Security (ICCAIS) (2019)
7. Kaur, M., Singh, D., Kumar, V., Gupta, B.B., Abd El-Latif, A.A.: Secure and energy efficient-based e-health care framework for green internet of things. IEEE Trans. Green Commun. Netw. **5**(3), 1223–1231 (2021)
8. Alanezi, A., Abd-El-Atty, B., Kolivand, H., Abd El-Latif, A., El-Rahiem, B., Sankar, S., Khalifa, H.: Securing digital images through simple permutation-substitution mechanism in cloud-based smart city environment. Secur. Commun. Netw. **2021**, 1–17 (2021)
9. Abd El-Latif, A., Abd-El-Atty, B., Elseuofi, S., Khalifa, H., Alghamdi, A., Polat, K., Amin, M.: Secret images transfer in cloud system based on investigating quantum walks in steganography approaches. Phys. A Statist. Mech. Appl. **541**, 123687 (2019)
10. Abd El-Latif, A.A., Abd-El-Atty, B., Mehmood, I., Muhammad, K., Venegas-Andraca, S.E., Peng, J.: Quantum-inspired blockchain-based cybersecurity: securing smart edge utilities in IoT-based smart cities. Inf. Process. Manag. **58**(4), 102549 (2021)
11. Abd-El-Atty, B., Iliyasu, A., Alaskar, H., Abd El-Latif, A.: A robust quasi-quantum walks-based steganography protocol for secure transmission of images on cloud-based e-healthcare platforms. Sensors **20**, 3108 (2020)
12. Abbas, K., Tawalbeh, L., Rafiq, A., Muthanna, A., Elgendy, I., Abd El-Latif, A.: Convergence of blockchain and iot for secure transportation systems in smart cities. Secur. Commun. Netw. **2021**, 1–13 (2021)

13. Nguyen, G.N., Viet, N.H.L., Elhoseny, M., Shankar, K., Gupta, B.B., Abd El-Latif, A.A.: Secure blockchain enabled cyber—physical systems in healthcare using deep belief network with resnet model. J. Parall. Distrib. Comput. **153**, 150–160 (2021)
14. Abou-Nassar, E.M., Iliyasu, A.M., El-Kafrawy, P.M., Song, O.-Y., Bashir, A.K., Abd El-Latif, A.A.: Ditrust chain: towards blockchain-based trust models for sustainable healthcare IoT systems. IEEE Access **8**, 111223–111238 (2020)
15. Tsafack, N., Sankar, S., Abd-El-Atty, B., Kengne, J., Jithin, K.C., Belazi, A., Mehmood, I., Bashir, A.K., Song, O.-Y., Abd El-Latif, A.A.: A new chaotic map with dynamic analysis and encryption application in internet of health things. IEEE Access **8**, 137731–137744 (2020)
16. Shi, W., Cao, J., Zhang, Q., Li, Y., Xu, L.: Edge computing: vision and challenges. IEEE Int. Things J. **3**(5), 637–646 (2016)
17. Armbrust, M., Fox, A., Griffith, R., Joseph,, A.D. Katz, R.H., Konwinski, A., Lee, G., Patterson, D.A., Rabkin, A., Stoica, I., Zaharia, M.: Above the clouds: A berkeley view of cloud computing. Technical Report UCB/EECS-2009-28, EECS Department, University of California, Berkeley (2009)
18. Zhang, Q., Cheng, L., Boutaba, R.: Cloud computing: state-of-the-art and research challenges. J. Int. Serv. Appl. **1**(1), 7–18 (2010)
19. Mell, P., Grance, T.: The NIST definition of cloud computing. Technical Report 800-145, National Institute of Standards and Technology (NIST), Gaithersburg, MD (2011)
20. Zhang, W.-Z., Elgendy, I.A., Hammad, M., Iliyasu, A.M., Du, X., Guizani, M., Abd El-Latif, A.A.: Secure and optimized load balancing for multitier iot and edge-cloud computing systems. IEEE Int. Things J. **8**(10), 8119–8132 (2021)
21. Asim, M., Wang, Y., Wang, K., Huang, P.-Q.: A review on computational intelligence techniques in cloud and edge computing. IEEE Trans. Emerg. Top. Comput. Intell. **4**(6), 742–763 (2020)
22. Asim, M., Mashwani, W.K., Shah, H., Belhaouari, S.B.: An evolutionary trajectory planning algorithm for multi-UAV-assisted MEC system. Soft Comput. **26**, 7479–7492 (2021)
23. Asim, M., Abd El-Latif, A.A.: Intelligent computational methods for multi-unmanned aerial vehicle-enabled autonomous mobile edge computing systems. ISA Trans. **132**, 5–15 (2021)
24. Abd El-Latif, A.A., Abd-El-Atty, B., Mazurczyk, W., Fung, C., Venegas-Andraca, S.E.: Secure data encryption based on quantum walks for 5g internet of things scenario. IEEE Trans. Netw. Serv. Manag. **17**(1), 118–131 (2020)
25. Rohan, R., Prasad, C.N., Jeswin, J., Sadiq, P.: An introduction to the collective behaviour of swarm intelligence. In: National Conference on Contemporary Research and Innovations in Computer Science (NCCRICS) (2017)
26. Granjal, J., Monteiro, E., Sa Silva, J.: Security for the internet of things: a survey of existing protocols and open research issues. IEEE Commun. Surv. Tutor. **17**(3), 1294–1312 (2015)
27. El-Latif, A., Ahmed, A., Tawalbeh, L.A.: New trends of intelligent systems-based secure critical infrastructure in smart city. J. Intell. Fuzzy Syst. 1–1 (2021)
28. Zhang, W.-Z., Elgendy, I.A., Hammad, M., Iliyasu, A.M., Du, X., Guizani, M., Abd El-Latif, A.A.: Secure and optimized load balancing for multitier iot and edge-cloud computing systems. IEEE Int. Things J. **8**(10), 8119–8132 (2021)
29. Elgendy, I.A., Yadav, R.: Survey on Mobile Edge-Cloud Computing: A Taxonomy on Computation offloading Approaches, pp. 117–158. Springer International Publishing, Cham, (2022)
30. Singh, S., Sulthana, R., Shewale, T., Chamola, V., Benslimane, A., Sikdar, B.: Machine learning assisted security and privacy provisioning for edge computing: A survey. IEEE Int. Things J. **9**, 1–1 (2021)
31. Lansky, J., Ali, S., Mohammadi, M., Majeed, M.K., Karim, S.H.T., Rashidi, S., Hosseinzadeh, M., Rahmani, A.M.: Deep learning-based intrusion detection systems: a systematic review. IEEE Access **9**, 101574–101599 (2021)
32. Zargar, S.T., Joshi, J., Tipper, D.: A survey of defense mechanisms against distributed denial of service (DDOS) flooding attacks. IEEE Commun. Surv. Tutor. **15**(4), 2046–2069 (2013)

33. Masdari, M., Ahmadzadeh, S.: Comprehensive analysis of the authentication methods in wireless body area networks: Authentication methods in wireless body area networks. Secur. Commun. Netw. **9**, 10 (2016)
34. Sun, Y., Lo, F.P.-W., Lo, B.: Security and privacy for the internet of medical things enabled healthcare systems: a survey. IEEE Access **7**, 183339–183355 (2019)
35. Khan, K., Mehmood, A., Khan, S., Khan, M.A., Iqbal, Z., Mashwani, W.K.: A survey on intrusion detection and prevention in wireless ad-hoc networks. J. Syst. Archit. **105**, 101701 (2020)
36. Masdari, M., Khezri, H.: A survey and taxonomy of the fuzzy signature-based intrusion detection systems. Appl. Soft Comput. **92**, 106301 (2020)
37. Zhang, C., Patras, P., Haddadi, H.: Deep learning in mobile and wireless networking: a survey. IEEE Commun. Surv. Tutor. **21**(3), 2224–2287 (2019)
38. Wang, X., Han, Y., Leung, V.C.M., Niyato, D., Yan, X., Chen, X.: Convergence of edge computing and deep learning: a comprehensive survey. IEEE Commun. Surv. Tutor. **22**(2), 869–904 (2020)
39. Yi, S., Qin, Z., Li, Q.: Security and privacy issues of fog computing: A survey. In: Xu, K., Zhu, H. (Eds.), Wireless Algorithms, Systems, and Applications, pp. 685–695. Springer International Publishing, Cham (2015)
40. Jafarian, T., Masdari, M., Ghaffari, A., Majidzadeh, K.: A survey and classification of the security anomaly detection mechanisms in software defined networks. Cluster Comput. **24**, 1–19 (2021)
41. Pang, G., Shen, C., Cao, L., Hengel, A.: Deep learning for anomaly detection: a review. ACM Comput. Surv. **54**, 1–38 (2021)
42. Taleb, T., Samdanis, K., Mada, B., Flinck, H., Dutta, S., Sabella, D.: On multi-access edge computing: a survey of the emerging 5G network edge cloud architecture and orchestration. IEEE Commun. Surv. Tutor. **19**(3), 1657–1681 (2017)
43. Porambage, P., Okwuibe, J., Liyanage, M., Ylianttila, M., Taleb, T.: Survey on multi-access edge computing for Internet of things realization. IEEE Communic. Surv. Tutor. **20**(4), 2961–2991 (2018)
44. Pham, Q.-V., Fang, F., Vu, H.-N., Le, M., Ding, Z., Le, L.B., Hwang, W.-J.: A survey of multi-access edge computing in 5G and beyond: fundamentals, technology integration, and state-of-the-art. CoRR **abs/1906.08452** (2019)
45. Mach, P., Becvar, Z.: Mobile edge computing: a survey on architecture and computation offloading. IEEE Commun. Surv. Tutor. **19**(3), 1628–1656 (2017)
46. Ahmed, A., Ahmed, E.: A survey on mobile edge computing. In: 2016 10th International Conference on Intelligent Systems and Control (ISCO), pp. 1–8 (2016)
47. Ren, J., Zhang, D., He, S., Zhang, Y., Li, T.: A survey on end-edge-cloud orchestrated network computing paradigms: transparent computing, mobile edge computing, fog computing, and cloudlet. ACM Comput. Surv. **52**(6), 125:1–125:36 (2019)
48. Wang, S., Zhang, X., Zhang, Y., Wang, L., Yang, J., Wang, W.: A survey on mobile edge networks: convergence of computing, caching and communications. IEEE Access **5**, 6757–6779 (2017)
49. Vaquero, L.M., Rodero-Merino, L.: Finding your way in the fog: towards a comprehensive definition of fog computing. SIGCOMM Comput. Commun. Rev. **44**(5), 27–32 (2014)
50. Shi, W., Cao, J., Zhang, Q., Li, Y., Xu, L.: Edge computing: vision and challenges. IEEE Int. Things J. **3**(5), 637–646 (2016)
51. Satyanarayanan, M., Bahl, P., Caceres, R., Davies, N.: The case for VM-based cloudlets in mobile computing. IEEE Pervas. Comput. **8**(4), 14–23 (2009)
52. Satyanarayanan, M., Simoens, P., Xiao, Y., Pillai, P., Chen, Z., Ha, K., Wenlu, H., Amos, B.: Edge analytics in the Internet of things. IEEE Pervas. Comput. **14**(2), 24–31 (2015)
53. Gao, Y., Hu, W., Ha, K., Amos, B., Pillai, P., Satyanarayanan, M.: Are cloudlets necessary? Technical Report, School of Computer Science, Carnegie Mellon University, Pittsburgh, PA, USA (2015)

54. Bonomi, F., Milito, R., Zhu, J., Addepalli, S.: Fog computing and its role in the Internet of things. In: Proceedings of the First Edition of the MCC Workshop on Mobile Cloud Computing, MCC '12, pp. 13–16. ACM, New York (2012)
55. Chiang, M.: Fog networking: an overview on research opportunities. CoRR **abs/1601.00835** (2016)
56. Tandon, R., Simeone, O.: Harnessing cloud and edge synergies: toward an information theory of fog radio access networks. IEEE Commun. Mag. **54**(8), 44–50 (2016)
57. Chiang, M., Zhang, T.: Fog and IoT: an overview of research opportunities. IEEE Int. Things J. **3**(6), 854–864 (2016)
58. Hu, W., Gao, Y., Ha, K., Wang, J., Amos, B., Chen, Z., Pillai, P., Satyanarayanan, M.: Quantifying the impact of edge computing on mobile applications. In: Proceedings of the 7th ACM SIGOPS Asia-Pacific Workshop on Systems, APSys '16, pp. 5:1–5:8. ACM, New York (2016)
59. Ni, J., Zhang, K., Lin, X., Shen, X.S.: Securing fog computing for Internet of things applications: challenges and solutions. IEEE Commun. Surv. Tutor. **20**(1), 601–628 (2018)
60. Chandola, V., Banerjee, A., Kumar, V.: Anomaly detection: a survey. ACM Comput. Surv. **41**(3), 58 (2009)
61. Lofti, A.: Zadeh. Fuzzy logic. Comput. **21**(4), 83–93 (1988)
62. Zadeh, L.A.: Fuzzy sets. Inf. Control **8**(3), 338–353 (1965)
63. Takagi, T., Sugeno, M.: Fuzzy identification of systems and its applications to modeling and control. IEEE Trans. Syst. Man Cyber. **SMC-15**, 116–132 (1985)
64. Cao, S.G., Rees, N.W., Feng, G.: Mamdani-type fuzzy controllers are universal fuzzy controllers. Fuzzy Sets Syst. **123**(3), 359–367 (2001)
65. Witten, I.H., Frank, E., Hall, M.A.: Data Mining: Practical Machine Learning Tools and Techniques. Morgan Kaufmann Series in Data Management Systems, 3rd edn. Morgan Kaufmann, Amsterdam (2011)
66. Hinton, G.E., Osindero, S., Teh, Y.: A fast learning algorithm for deep belief nets. Neur. Comput. **18**(7), 1527–1554 (2006)
67. Mao, Q., Hu, F., Hao, Q.: Deep learning for intelligent wireless networks: a comprehensive survey. IEEE Commun. Surv. Tutor. **20**(4), 2595–2621 (2018)
68. Mohammadi, M., Al-Fuqaha, A., Sorour, S., Guizani, M.: Deep learning for IoT big data and streaming analytics: a survey. IEEE Commun. Surv. Tutor. **20**(4), 2923–2960 (2018)
69. Sutton, R.S., Barto, A.G.: Introduction to Reinforcement Learning, 1st edn. MIT Press, Cambridge (1998)
70. Shoham, Y., Powers, R., Grenager, T.: Multi-agent reinforcement learning: a critical survey, technical report. Technical Report, Computer Science Department, Stanford University (2003)
71. François-Lavet, V., Henderson, P., Islam, R., Bellemare, M.G., Pineau, J.: An introduction to deep reinforcement learning. CoRR **abs/1811.12560** (2018)
72. Li, G., Zhou, H., Feng, B., Li, G., Li, T., Xu, Q., Quan, W.: Fuzzy theory based security service chaining for sustainable mobile-edge computing. Mob. Inf. Syst. **2017**, 8098394:1–8098394:13 (2017)
73. Ritu, Jain, S.: A trust model in cloud computing based on fuzzy logic. In: IEEE International Conference On Recent Trends In Electronics Information Communication Technology, pp. 59–66 (2016)
74. Qu, C., Buyya, R.: A cloud trust evaluation system using hierarchical fuzzy inference system for service selection. In: 2014 IEEE 28th International Conference on Advanced Information Networking and Applications. IEEE, Piscataway (2014)
75. Balan, E.V., Priyan, M.K., Gokulnath, C., Usha Devi, G.: Fuzzy based intrusion detection systems in manet. Procedia Comput. Sci. **50**, 109–114 (2015). Big Data, Cloud and Computing Challenges
76. Devi, R., Jha, R.K., Gupta, A., Jain, S., Kumar, P.: Implementation of intrusion detection system using adaptive neuro-fuzzy inference system for 5g wireless communication network. AEU- Int. J. Electron. Commun. **74**, 94–106 (2017)

77. Mansour, M.B., Abdelkader, T., Hashem, M., El-Horbaty, E.S.M.: An integrated three-tier trust management framework in mobile edge computing using fuzzy logic. PeerJ Comput. Sci. **7**, e700 (2021)
78. Chen, Y., Zhang, Y., Maharjan, S.: Deep learning for secure mobile edge computing. CoRR **abs/1709.08025** (2017)
79. Zhang, J., Guo, H., Liu, J.: A reinforcement learning based task offloading scheme for vehicular edge computing network. In: Han, S., Ye, L., Meng, W. (eds.) Artificial Intelligence for Communications and Networks, pp. 438–449. Springer International Publishing, Cham (2019)
80. Jabbar, W.A., Ahmed, A.A., Ali Safaa, S., Hiran, P.: Deep learning-based classification model for botnet attack detection. J. Amb. Intell. Humaniz. Comput. **13**, 3457–3466 (2022)
81. Wang, X., Zhou, Q., Harer, J., Brown, G., Qiu, S., Dou, Z., Wang, J., Hinton, A., Gonzalez, C.A., Chin, P.: Deep learning-based classification and anomaly detection of side-channel signals. In: Ternovskiy, I.V., Chin, P. (eds.) Cyber Sensing 2018, vol. 10630, pp. 37–44. International Society for Optics and Photonics, SPIE, Bellingham (2018)
82. Hussain, B., Du, Q., Zhang, S., Imran, A., Imran, M.A.: Mobile edge computing-based data-driven deep learning framework for anomaly detection. IEEE Access **7**, 137656–137667 (2019)
83. Chen, Y., Zhang, Y., Maharjan, S., Alam, M., Ting, W.: Deep learning for secure mobile edge computing in cyber-physical transportation systems. IEEE Netw. **33**(4), 36–41 (2019)
84. Hussain, B., Du, Q., Ren, P.: Deep learning-based big data-assisted anomaly detection in cellular networks. In: 2018 IEEE Global Communications Conference (GLOBECOM), pp. 1–6 (2018)
85. Masood, U., Asghar, A., Imran, A., Mian, A.N.: Deep learning based detection of sleeping cells in next generation cellular networks. In: 2018 IEEE Global Communications Conference (GLOBECOM), pp. 206–212 (2018)
86. Chi, J., Owusu, E., Yin, X., Yu, T., Chan, W., Liu, Y., Liu, H., Chen, J., Sim, S., Iyengar, V., Tague, P., Tian, Y.: Privacy partition: A privacy-preserving framework for deep neural networks in edge networks. In: 2018 IEEE/ACM Symposium on Edge Computing (SEC), pp. 378–380 (2018)
87. Nagisetty, A., Gupta, G.P.: Framework for detection of malicious activities in IoT networks using keras deep learning library. In: 2019 3rd International Conference on Computing Methodologies and Communication (ICCMC), pp. 633–637 (2019)
88. Huang, B., Li, Y., Li, Z., Pan, L., Wang, S., Xu, Y., Hu, H.: Security and cost-aware computation offloading via deep reinforcement learning in mobile edge computing. Wirel. Commun. Mob. Comput. **2019**, 20 (2019)
89. Gupta, B.B., Varun, P., Nadia, N., Vijayakumar, P., Abd El-Latif, A.A., Xiaojun, C.: Machine learning and smart card based two-factor authentication scheme for preserving anonymity in telecare medical information system (TMIS). Neur. Comput. Appl. **35**, 5055–5080 (2023)
90. Huang, B., Xiang, Y., Yu, D., Wang, J., Li, Z., Wang, S.: Reinforcement learning for security-aware workflow application scheduling in mobile edge computing. Secur. Commun. Netw. **2021**, 13 (2021)
91. Sedik, A., Hammad, M., Abd El-Latif, A.A., El-Banby, G.M., Khalaf, A.A., Abd El-Samie, F.E., Iliyasu, A.M.: Deep learning modalities for biometric alteration detection in 5g networks-based secure smart cities. IEEE Access **9**, 94780–94788 (2021)
92. Xu, M., Peng, J., Gupta, B.B., Kang, J., Xiong, Z., Li, Z., El-Latif, A.A.A.: Multi-agent federated reinforcement learning for secure incentive mechanism in intelligent cyber-physical systems. IEEE Int. Things J. **9**, 22095–22108 (2021)
93. Parizi, R.M., Pouriyeh, S., Dehghantanha, A., Mothukuri, V., Khare, P., Srivastava, G.: Federated learning-based anomaly detection for iot security attacks. IEEE Int. Things J. **9**, 1–1 (2021).
94. Mighan, S.N., Kahani, M.: A novel scalable intrusion detection system based on deep learning. Int. J. Inf. Secur. **20**, 06 (2021)
95. Popoola, S.I., Adebisi, B., Hammoudeh, M., Gacanin, H., Gui, G.: Stacked recurrent neural network for botnet detection in smart homes. Comput. Electri. Eng. **92**, 107039 (2021)

Machine Learning Techniques for Secure Edge SDN

Yassine Maleh, Abdelkebir Sahid, Ahmed A. Abd El-Latif, and Karim Ouazzane

1 Introduction

Today, software-defined network (SDN), network functions virtualization (NFV), and software-defined wide-area network (SD-WAN) approaches are disrupting traditional network systems. They are initiating a "real change" in enterprise network infrastructures with promises of cost reduction, agility, flexibility, speed of implementation, and adaptability. The trends follow one another and have a strong impact on network design itself. SDN, NFV, and SD-WAN have become the basic building blocks for preparing your network for the challenges of today and tomorrow. SDN and NFV are similar in many ways; they have the same components and pursue similar goals. In the network functions virtualization (NFV) approach, network functions based on physical appliances are virtualized via a hypervisor, allowing the network to grow without needing dedicated devices. This solution separates network control functions such as routers, firewalls, and load balancers

Y. Maleh (✉)
Laboratory of Sciences and Techniques for Engineering (LaSTI), ENSAK, Sultan Moulay Slimane University (USMS), Beni Mellal, Morocco

A. Sahid
University Hassan 1st, Settat, Morocco

A. A. Abd El-Latif
Center of Excellence in Quantum and Intelligent Computing, Prince Sultan University, Riyadh, Saudi Arabia

Department of Mathematics and Computer Science, Faculty of Science, Menoufia University, Shebin El-Koom, Egypt
e-mail: aabdellatif@psu.edu.sa

K. Ouazzane
Cyber Security Research Centre, London Metropolitan University, London, UK

A. A. Abd El-Latif et al. (eds.), *Secure Edge and Fog Computing Enabled AI for IoT and Smart Cities*, EAI/Springer Innovations in Communication and Computing,
https://doi.org/10.1007/978-3-031-51097-7_14

from the hardware layer and allows these more advanced features to run from a virtual machine.

A single hypervisor orchestrates all virtual machines. When the hypervisor controls the network functions, the services can be run on standard servers [19]. These software modules can interact to provide communication services previously handled by hardware platforms. SDN and NFV are generally independent, although NFV can greatly complement SDN. The growth that data center networks have shown has turned these networks into a key element of the corporate IT infrastructure. For data center networks to keep up with the growth of data from new sources, new server capabilities are required for storage and handling increasingly complex service interactions. However, the architectures and technologies of traditional data centers are no longer keeping up with the emerging demands for high efficiency, intelligence, and usability. This has led to the need for SDN.

Through open protocols such as OpenFlow, SDN designs may make network management programmable. As a result, companies may implement edge-based software control that is aware of their networks. Instead of utilizing closed and proprietary firmware to setup, manage, protect, and optimize network resources, this gives users access to network switches and routers [32].

SDN solutions allow new components to be added or existing ones to be configured through programmable interfaces or scripts. Administrators can now dynamically and automatically adjust traffic flows across the network to meet changing needs. Network intelligence is centralized through SDN controllers implemented in the control software. This control software provides a consistent, global view of the network. Applications that access geographically dispersed data and services across the public and private cloud require flexible traffic management. SDN enables this flexibility by allowing the bandwidth to scale as traffic changes. SDN is based on separating the network infrastructure from the physical devices. By applying scheduling and automation to network control, network operators can dynamically define and manage their networks. However, this innovative and improved technology also introduces new security challenges to the network architecture. The central controller is now the main target of attacks. The controller is vulnerable to various attacks, such as distributed denial-of-service (DDoS) attacks, which can deplete network resources [63]. As a result, the services provided by the controller can become unavailable. Integrating an intrusion detection system (IDS) is necessary to protect SDN. Attack detection requires an adaptive and accurate classifier to make decisions based on uncertain information. It is very important to detect an attack on a controller early. An important feature of an SDN network is that OpenFlow technology allows you to use the network as a sensor, because information about the flows can be obtained from all critical points. We address these challenges by answering the following research question in this chapter:

- RQ1: What are the main security threats in SDN?
- RQ2: What are the latest studies on IDS-based machine learning in SDN?
- RQ3: What are the challenges of using IDS-based machine learning in SDN?

This chapter discusses the main advantages of SDN, possible types of attacks on this type of network, and current work on securing SDN using methods based on machine learning. This chapter is organized as follows: Section 2 briefly describes the SDN architecture and the main benefits of SDN. Section 3 discusses security threats in SDN. Then Sect. 4 mitigates attacks in SDN, particularly on intrusion detection techniques using machine learning techniques. Section 5 presents the chapter's conclusion and some future directions.

2 Background

2.1 SDN Architecture

A software-defined network is a method of computer network administration that allows network services to be managed when the control plane functionality is separated (abstracted) from the underlying packet forwarding layer (data plane). Network planning and traffic control are handled programmatically. Application programming interfaces (APIs) are provided for the upper layer applications. In this way, the introduction of new services on the network is accelerated and facilitated. Three layers are distinguished in the SDN architecture, as described in Fig. 1.

1. The infrastructure layer includes a set of network devices [52].
2. The control layer includes an operating system that provides network services to applications and a software interface for managing network devices and the network.
3. The application layer manages the network.

Software-defined networks are often used to build cloud infrastructure services where, at the request of service consumers, it is necessary to create virtual nodes and allocate virtual network resources automatically and in the shortest possible time. Also, the use of software-defined networks is appropriate in large data centers, reducing network maintenance costs due to the centralization of management on the software controller and increasing the percentage of network resources through dynamic management [39]. The most promising and actively developing protocol for SDN is OpenFlow. Still, there are others: ForCES, Open vSwitch database (OVSDB), POF OpFlex, OpenState, revised open-flow library (ROFL), hardware abstraction layer (HAL), and programmable abstraction of data path. OpenFlow is a protocol for controlling the processing of data transmitted over a data network by routers and switches that implements the technology of software-defined networks. The first version of OpenFlow (v1.1) was implemented on February 28, 2011 [42]. The standard was developed by the Open Networking Foundation (ONF). The protocol controls network switches and routers from a central device, called a network controller. The controller in a network can be a separate server or even an ordinary administrator's computer, on which a network operating system

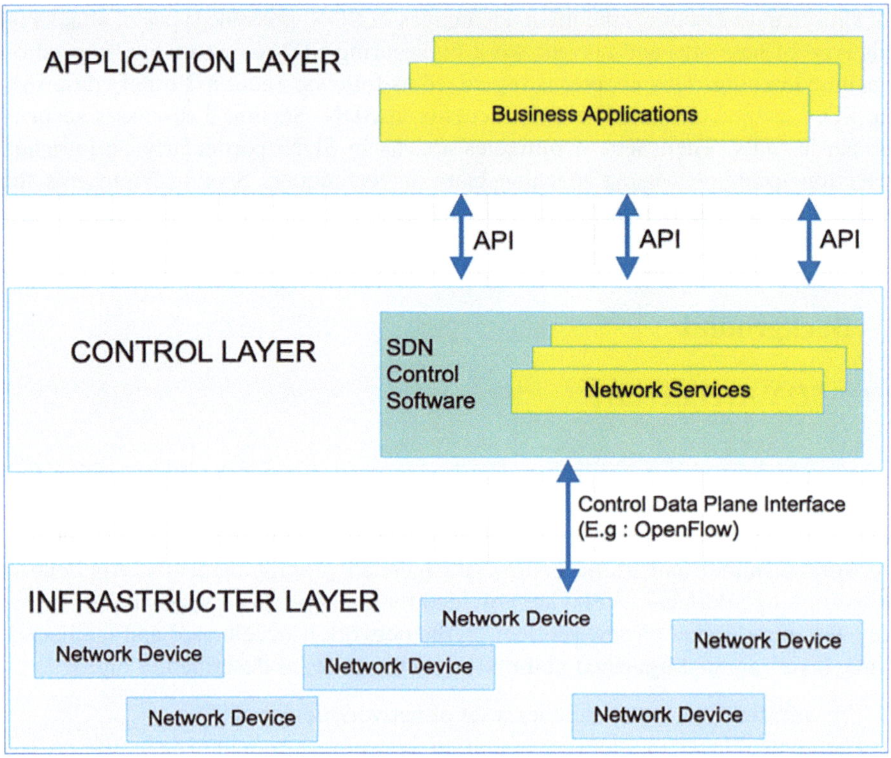

Fig. 1 SDN architecture

is installed, providing management interfaces between network applications and network switches. The controller is used to manage switch flow tables, based on which the decision is made to transfer the received packet to a particular switch port. This way, direct network connections are formed within the network with minimum data transfer delays and the required parameters.

2.2 SDN Benefits

As a result of the rapid development of information technology, we can observe the following trends in the development of corporate networks and data center networks: Rapid growth in traffic volumes and a change in its structure toward video transmission and unified communications. The need to process big data has led to the emergence of high-performance computing centers.

The way telecom businesses do business is changing as a result of SDN. Verizon, for example, utilizes SDN to consolidate all of its Ethernet and IP-based service edge routers onto a single platform. Improving operational efficiency and flexibility for new activities and services is simplifying the edge architecture.

The problem with traditional large-scale networks is the distributed management of network devices [52]. Changing the configuration of such a network is complex and can be time-consuming. SDN can provide new opportunities, allowing us to solve the problems of increasing bandwidth, simplifying network management, load redistribution, and increasing network scalability [76]. The main advantages of SDN architecture over traditional, modern networks are as follows:

- Simplified Policy Changes.

  It is now possible for an administrator to fine-tune any network switch's rules, prioritizing or de-prioritizing packets, and even banning particular kinds of packets with SDN.

  Administrators benefit greatly from this feature when working with a multi-tenant cloud computing architecture since it allows them to control traffic loads better. Administrators may now employ less expensive commodity switches and have more control over how network traffic is routed.
- Network Management and Visibility.

  SDN also provides network control and end-to-end visibility as additional advantages. A network administrator only needs to deal with one centralized controller to deliver rules to linked switches. Contrast this with setting up and configuring many devices at one time.

  The network controller's capacity to monitor traffic and implement security controls is also a benefit in terms of security. In this context, traffic that the controller deems suspicious might be rerouted or dropped.
- Reduced Hardware Footprint and Opex.

  SDN also enables the virtualization of hardware and services previously managed by dedicated hardware. This results in a smaller hardware footprint and lower operational expenses.
- Networking Innovations.

  SDN also influences software-defined wide-area network (SD-WAN) technology. SDN's virtual overlay is utilized in SD-WAN. As a virtual network that may use whatever connection the controller deems appropriate to transport traffic, SD-WAN abstracts an organization's connectivity connections across its WAN and creates an SD-WAN virtual network.

3 Security Threats in SDN

Because of the separation of the control and data planes, SDNs are more programmable, scalable, and adaptable than their conventional counterparts. However, this has resulted in additional security concerns specific to the SDN. Although many cloud and network service providers appear to have been persuaded by the idea's numerous advantages, it cannot be broadly used unless its security issues are solved.

Because the management subsystem is located outside, it is the most vulnerable location in the system to targeted assaults that aim to disrupt the SDN software controller [3]. A controller failure would bring the entire network to a screeching halt if each packet was routed independently, and a single control center was no longer vulnerable. Attackers are drawn to such a location. Switches will likely go offline in route calculation mode and resume the old technology if the central control fails, although this would not damage the link. As a result, it is doubtful that device designers would entirely ditch support for tried and true technologies and instead choose "pure" SDN/OpenFlow. An assault on SDN is broader than an attack on IP networks because it includes a wide range of targets: the controller, communication routes between the controller and switch, and switches themselves.

In the IP protocol, routers track how much traffic flows through each device's channels and how well it performs. The controller in SDN keeps tabs on the network's status by receiving data on the failure of channels and the load on those channels "from outside." Delays in getting information and making judgments are detrimental to trustworthiness. Multiple instructions sent to devices that already support OpenFlow might cause problems for the network or a specific application since the network management protocol does not presuppose "intelligence" in the switch. Network traffic–interception devices controlled by intruders can be installed using this protocol. Companies must keep an eye out for the OpenFlow protocol in their network, as hardware manufacturers have already begun to include support for it in their products. As a result, users who had not intended to create their SDN may be subject to the unlawful usage of OpenFlow. Another difficulty with SDN is the ability to change the network to match the needs of a certain application or virtualization management system by executing programs on the controller. These applications have yet to be tested in terms of real-time competition. Creating these network applications is challenging, and cloud users may not do so to their full potential. So, it is conceivable that certain programs will dominate network resources and essentially shut down all other applications at some time in the future. The network optimization technique might be implemented wrongly in a second way by application developers. Tracking this issue is challenging since it only arises under specific circumstances and heavy loads. Configuration problems should theoretically be recognized by network management software, but this requires that software vendors support SDN technology. In any case, the data center operator must keep an eye on user applications and prevent them from monopolizing the network.

To adopt proactive security measures, these issues must be brought to the attention of policymakers and the general public. Many authors have analyzed the security of the SDN [20, 68]. In Fig. 2, there are several risks and vulnerabilities at the various levels and interfaces of the SDN network. All SDN layers and interfaces are at risk in this area.

Managing a network through a controller is essential for the Internet of Things and other future networks. However, it also raises additional security concerns that need to be considered. All kinds of problems arise, from crucial points in SDN controllers to the necessity for identification and permission of network

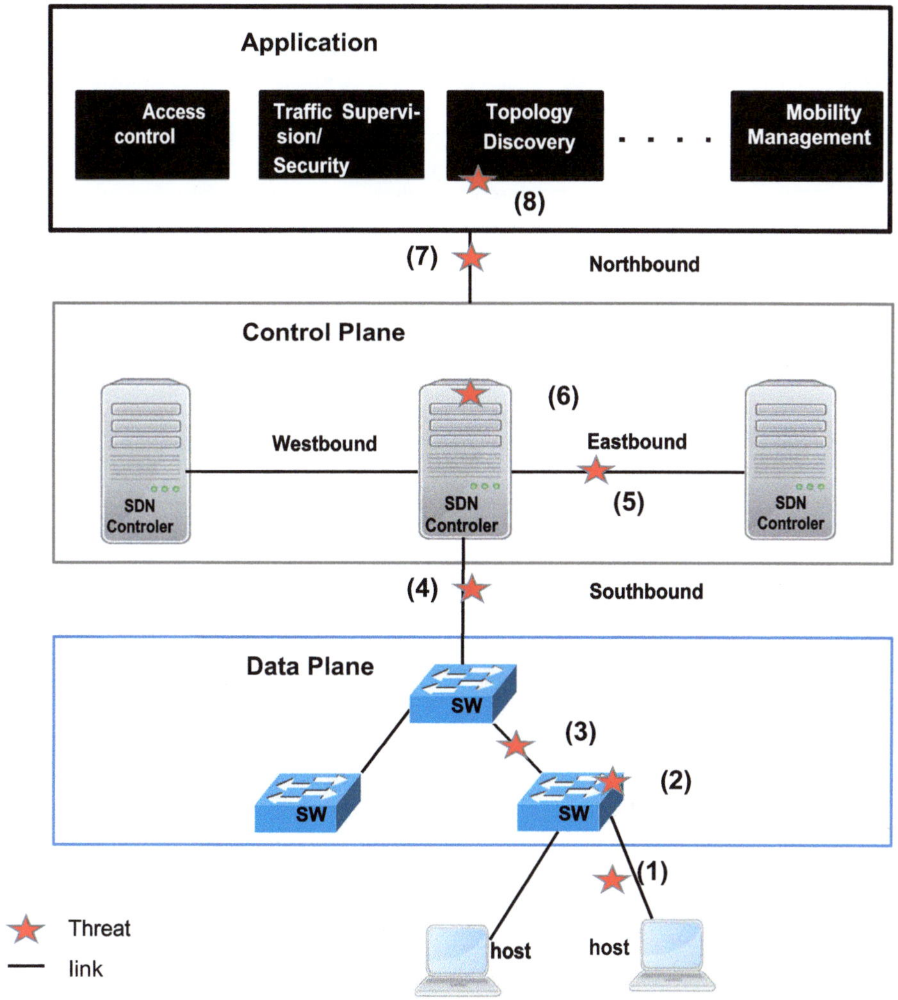

Fig. 2 SDN's threat analysis

applications. Furthermore, as the SDN is gradually implemented, the number of security issues is projected to grow. Table 1 summarizes the primary security vulnerabilities in the three SDN levels.

Because DoS is an assault, the selected research focused on the attack's weaknesses rather than the attack itself. The papers [20, 68] discussing DoS and DDoS assaults were all collected in one place. As a result of resource exhaustion, controllers and switches can no longer operate at full capacity (CPU and memory usage). With packet flooding and the difficulties of detecting actual devices in the network, growing the network's capacity with phony devices makes network administration impractical. Vulnerabilities in the access control subcategory take

Table 1 Major security threats in SDN

SDN plane/layer	Proposal	Threat type	Description
Application plane	[18, 47, 50, 69, 72, 74]	Lack of authentication and authorization	Applications need a reliable method of establishing trust and granting access.
Control plane	[43, 57, 73]	Fraudulent insertion of flow rules	Malicious or hacked programs can generate false flow rules, and determining if they result from a compromised application is complex.
	[26, 31, 33, 69, 53]	Lack of access control and traceability	Third-party apps are challenging to establish access control and traceability.
	[5, 7, 9, 13, 16, 23, 28, 75]	Denial-of-service attacks (DoS)	Attackers use denial-of-service (DoS) techniques to overwhelm a network controller's processing capacity.
Data plane	[14, 25, 34]	Unauthorized access to the controller	The program lacks a robust way to provide access control.
	[11, 77, 78]	Interception	Only a finite number of flow rules may be stored in OpenFlow switch flow tables.
	[8, 27, 37, 41, 54, 62]	Misconfiguration	OpenFlow switches' flow tables can only hold so many flow rules.
	[12, 15, 59, 79]	Controller compromise	The security of the data plan is dependent on the implementation of a control plan. The controller's security affects the data plan's security.
	[1, 4, 21, 40, 44, 45, 49, 65, 67]	Man-in-the-middle attack	Due to the optional use and the complexity of TLS protocol.

advantage of the fact that SDN components can be accessed via issues with access authorization policies and delegations when exploited. The authentication category is vulnerable to the impersonation problem, which is the difficulty in determining if an entity in the network is itself (be it a controller device, switch, or even network administrator). Studies involving unauthorized access to private data are under the purview of the confidentiality breach category. A vulnerability in this category can lead to the hijacking of information such as flow tables and flow rules. When network managers purposefully or unintentionally introduce misconfigurations into the network infrastructure, they fall under the heading "misconfiguration." For example, integrity violation explores the potential of putting faked data into the SDN network, where the actual data, for example, packet data and flow rules, are modified. Problems resulting from a faulty controller are included under the term "controller failure [30, 36]." This vulnerability group aims to learn more about system failure's underlying causes and effects. The interval between network configuration changes, such as flow rules or firmware, is the topic of the network update category. Conflicts and, thus, security flaws might emerge if a device relies on the device's configuration being updated within this time.

As a result, the lack of TLS category studies reveals the consequences of not employing TLS to secure communication between SDN devices [61]. Although the TLS protocol may be a solution, network equipment makers' implementation of TLS has been criticized. The OpenFlow protocol, for example, has a poor definition in this aspect [44]. However, it does not discuss how TLS should be used in device-to-device communication. SDN attacks are rare because of the lack of research into security in SDN. Table 1 shows some of the assaults documented in the literature. When an attacker repeatedly injects requests for flows into the connection between the controller and switch, they commit control plane saturation and interception, where they collect control plane traffic to redirect, examine, or alter it.

Moreover, poisoning and flow table overflow attacks were discovered; in both cases, the attacker injects packets into the network, leading to a denial of service. In poisoning attacks, the objective is to cause a denial of service by creating new flows that exceed the capacity of flow tables. A user's account can also be compromised through a man-in-the-middle (MitM) attack. Finally, the fingerprinting attack uses variations in network metrics like the time difference when routing packets across the network to investigate information about the controller, routes, or essential devices on the network, such as servers.

4 Machine Learning for Secure SDN

We discuss how research over the past 5 years has shown that the integration of ML and SDN has been crucial in detecting intrusions. Methods for detecting intrusions in SDN using machine learning are listed in Table 2.

The challenge of protecting SDN involves automating the detection of possible threats and the subsequent formulation of policy. Defending against several assaults at once may be impossible in real time, even if a network administrator can pinpoint each attack and its source. Therefore, the SDN controller must apply specific security rules similar to firewall rules. Defining these rules might be difficult since they should restrict access for malicious hosts and attackers while leaving regular users unrestricted. Various methods, including machine learning, can identify malicious users in the user flow. However, machine learning approaches promise to solve this problem [58]. A multivector DDoS attack detection system based on deep learning is proposed for a software-defined network [51]. The authors implemented the system as a network application on top of an SDN controller. Deep learning reduces the number of features from a large set derived from network traffic headers. The system is evaluated based on various metrics (accuracy, precision, recall, f-measure, Receiver Operating Characteristic (ROC).

Table 2 SDN machine learning methods based on IDS

Paper	Algorithms	Targeted threat	Controller	Dataset	Accuracy
[70]	C4.5 DT	Botnet	Opendaylight	CTU-13, ISOT	97%, 90%
[10]	Statistical method + machine learning	DDoS	Floodlight	UNB-ISCX, CTU-13, and ISOT	99.12%
[46]	ASVM	Flooding-based DDoS	OpenDaylight	SDN traffic	97%
[29]	SVM	Probe, DoS, R2L, U2L, data attacks	Ryu	DARPA	84, 21%
[60]	SVM, NB, NC	–	POX	Traffic generated in SDN	92,3%; 96,79%; 91,02%
[66]	SVM, selective logging, IP traceback	–	Ryu	NSL-KDD	95,98%
[66]	SVM, selective logging, IP traceback	–	Ryu	9 features of NSL-KDD	87,74%
[6]	SVC-RF	DDoS	Ryu	Traffic generated in SDN	98,8%
[17]	Decision tree, random Forest, XGBoost	DDoS, PROBE, R2L and U2R	–	NSL-KDD	95,95%
[64]	SVM, KPCA and GA	DDoS	POX	ASNM-NPBO	98.907%
[55]	Q-learning	DoS	ONOS	Traffic generated in SDN	–

Nanda et al. [48] consider four well-known machine learning algorithms: decision trees (C4.5), Bayesian networks, decision tables, and naive Bayesian classifiers (naive Bayes, NB). The algorithms, relying on the collected "historical" data and the current behavior of the network traffic, must predict a possible attack on a particular node. Experimental results show that the Bayesian network algorithm showed the best prediction accuracy. The Bayesian network is a graph-based probabilistic model. A Bayesian network is a directed acyclic graph, each vertex corresponding to a random variable, and its arcs encode relations of conditional independence between these variables. The vertices can represent variables of any type, be they weighted parameters, latent variables, or hypotheses. With proper training, the Bayesian network can be used as a classifier and show accurate classification results. Another widely used classifier is the support vector machine (SVM). This method also offers fairly high accuracy and few false positives. Nokila et al. [35] presented an analysis of the SVM classifier and compared its effectiveness with other classifiers (RBF, naive Bayes, J48, and random forest). According to the experimental results, SVM showed the best accuracy and the lowest number of false positives. It is also possible to use deep neural networks for intrusion detection in SDN. Tang et al. [71] use gated recurrent unit recurrent neural network (GRU-RNN). The proposed approach is tested using the NSL-KDD dataset.

The experimental results show a high detection accuracy of 89%. The authors concluded that the proposed GRU-RNN operates efficiently without degrading the network performance. A multilayer perceptron can also be used to solve this problem. For example, [71] used a model based on a neural network trained on the NSL-KDD dataset. Then there are three hidden layers (12, 6, and 3 neurons). The output of the network has two neurons. This method showed good potential and, even on six features, showed accuracy (76%), comparable to the known machine learning algorithms (J48, naive Bayes (NB), NB tree, random forest, random tree, SVM). It follows that neural networks can detect abnormal traffic behavior in SDN. Even simple neural networks (e.g., a single-layer perceptron) can solve the problem of intrusion detection in software-defined networks [2]. The model was trained on the same NSL-KDD dataset. The backpropagation method is used to train the network. This model was able to show high accuracy on the test sample. This confirms the possibility and relevance of applying neural networks to solve the problem of intrusion detection in SDN. Li et al. [38] presented a DDoS detection model and defense system based on deep machine learning in a software-defined network environment. The performance of several neural networks (Long Short-Term Memory (LSTM), Convolutional Neural Network (CNN)/LSTM, GRU, 3LSTM) was compared. The accuracy was measured on the ISCX dataset. The 3LSTM model, 99%, showed the best result. This method compensates for the drawbacks of existing DDoS attack detection schemes. In [22], a fuzzy logic–based prototype intrusion detection system for SDN has been developed. The prototype consists of two modules: the statistics collection and processing module, and the decision-making module.

With the SDN-based cloud infrastructure, Phan et al. [56] suggested an effective technique to counteract DDoS attacks. The authors created a new machine learning model for traffic classification using a support vector machine and a self-organizing map method. To speed up and improve the accuracy of attack detection, the model uses an enhanced version of a history-based IP filtering technique (eHIPF). As a conclusion, the authors present a novel approach to developing a DDoS assault defender for the SDN-based cloud environment by combining the hybrid machine learning model with the eHIPF scheme. These approaches are optimal in terms of speed and simplicity of analysis. However, this approach has a few flaws that should be considered. The created dataset is quite limited, making it difficult to draw valid inferences from it (the sample size is too low). The model incorporates an upgraded history-based IP filtering method (eHIPF) to increase the attack detection rate and speed. Finally, the authors provide a unique technique that integrates the hybrid machine learning model and the eHIPF scheme to develop a DDoS assault defender for the SDN-based cloud environment. In terms of speed and ease of analysis, it is evident that these methods are ideal. There are, however, several drawbacks to this method. To begin with, the dataset generated is quite small, making it impossible to derive reliable conclusions from it (the number of samples is a few hundred or a few thousand in most cases). Moreover, the produced data comprises just a few types of assaults. These assaults do not realistically depict the diversity of anomalies found on the Internet. The attacker can readily replicate the detection mechanism's behavior by limiting the number of attack variants available. To begin with, the number of characteristics that can be retrieved is tiny compared to the total number of assaults. Furthermore, the training model's characteristics were mostly derived from the header packet, with no additional examination of the payload data. The header field may be readily rearranged to resemble other fields. For example, root to local (R2L) assaults and worm infestations can readily be fooled by malicious code in the payload packets. When it comes to application-level assaults, they are not very accurate. Tariq et al. [24] suggested a flow-based approach that uses ML algorithms to detect botnets in an SDN setting without inspecting the packet payload. In this study, the author combines real-time flow trace data with historical context to derive a better feature set, improving classification and reducing the false positive rate. A supervised machine learning technique based on a C4.5 decision tree is used to classify the botnets. The authors rely on publicly accessible dataset CTU-13 and ISOT botnets records to get botnet traces. Preliminary results show that the system can recognize known botnets with 97% in terms of accuracy and unknown botnets with 90% of accuracy. DDoS assaults can be detected using a methodology proposed in [10]. There were two stages to their task. An initial entropy measurement determines the distribution of each key characteristic extracted from the incoming packet traffic. A second entropy measurement follows this to determine the distribution of the other important features extracted. In the second step, the model uses the support vector machine (SVM) algorithm to predict whether or not the traffic is abnormal and therefore detect the DDoS attack. This is done after the entropy of the resulting features. Experiments using the SVM method demonstrated a 97% accuracy rate after assessment, as opposed to 91% accuracy rates with naive Bayes and 74% with random forest algorithms, respectively. It was claimed by [29, 46] that a new support vector machine (SVM) method for identifying DDoS assaults might be improved by using an advanced support vector machine (ASVM). A three-class classification method, ASVM, uses a multiclass approach. As a way to gauge the quality of the researchers' work, they employed three controllers instead of the more common usage of a single controller in previous SDN security studies. As a result of the attack, one controller is no longer functional, while the other controller is still operable. The researchers successfully detected the DDoS assaults based on flooding. The assessments considered false alarm rate, detection rate, and accuracy. A detection accuracy of around 97% was achieved with the quickest training and testing time. According to [29], flow-based IDS and packet-based IDS may be combined to achieve a high detection rate without affecting network performance. DARPA's intrusion detection dataset has been used to train support vector machines (SVMs) in the flow-based IDS. The initial layer of protection can identify any network intrusions. When an attack is discovered, the malicious flow is mirrored to a packet-based IDS for additional investigation and response. The evaluation findings revealed a detection rate of 84.21% and a good performance with little extra overhead. The authors proposed an algorithm based on three supervised learning models (SVM, closest centroid, and naive Bayes) [60]. (NB). They wanted to identify data traffic in the SDN environment depending on applications. Classifiers use network traffic traces and flow characteristics to predict future network traffic flow patterns. According to the authors, this study had a 92.3% success rate with SVM and a 96.7% success rate with nearest centroid. Support vector machines (SVMs) and selective logging for IP tracing were used [66] to create an SDN-based intrusion detection system (IDS). On selected aspects of the dataset, the assessment achieved 95.98% in terms of accuracy and 87.74% in terms of accuracy. For detection analysis, the authors say that SDN's centralized structure and the SVM logging's reasonable accuracy make this technique their first option. When an attack occurs, IP tracing may be used to identify the source of the packets, saving up to 95% of memory resources. Support vector classifier (SVC) outputs are used to classify traffic, which is then filtered by a random forest for distributed denial-of-service (DDoS) attack detection [6]. It has a success rate of 98.8%. The model has a sensitivity of 97.91% and an accuracy of 98.27%, which indicates that it accurately predicts the class label in most circumstances. By combining traditional and advanced tree-based ML algorithms, decision tree, random forest, and XGBoost, Chouhan et al. [17] introduced a method of traffic monitoring to identify hostile network activity as part of IDS in the SDN controller. The NSL-KDD dataset is the premier resource for developing and evaluating innovative IDS techniques. Extensive and intricate data preparation methods are employed to obtain the most reliable results. Only five NSL-41 KDD's characteristics were used to detect and categorize the attack (DDoS, PROBE, R2L, and U2R) with 95.95% accuracy. It is hypothesized by [64] that an SDN system may be used to identify and mitigate DDoS attacks. As a classifier, this model utilizes the SVM approach. Using the kernel principal component analysis (KPCA) and genetic algorithm (GA) techniques, the authors improved the model's accuracy and minimized the testing time. Data from the NSL-KDD set and another devoted to DDoS assaults were used in the experiment. The authors claim that KPCA extracts the most important features from the dataset, whereas GA is used to optimize SVM parameters by selecting the most relevant features for each SVM. Experiments reveal that KPCA works well on the DDoS dataset, and that the suggested model's accuracy is 99.997%. According to [55], machine learning has been used to construct a defensive architecture dubbed Q-MIND. They wanted to build a system to detect and respond to covert DoS attacks in SDNs. This method aims to improve detection accuracy using reinforcement learning based on Q-learning. Their Q-MIND techniques may identify and mitigate stealth DoS assaults more effectively than existing methods by employing the optimal Q-learning agent strategy, according to performance assessment findings.

5 Conclusion and Future Directions

In this overview study, we first outline the SDN architecture and how it contributes to the traditional or dispersed design's lack of manageability and programmability. It is easier than ever to administer a network using SDN, which separates the control plane from the data plane and allows for greater flexibility and scalability. As we have previously discussed, SDN-enabled networks are additionally vulnerable because of the separation of the control plane from the data plane. For this reason, scientists are working to find ways to protect the SDN network from attacks and breaches. If you have an infrastructure like this, the most effective technique for detecting abnormal behavior and alerting the network administrator is the intrusion detection system. It can classify network traffic and identify suspicious activity or patterns. An intrusion detection system depends on detecting anomalies quickly, accurately, and reliably. We have focused on machine learning–based IDS because we believe these critical factors are critical to an IDS's success. In addition, we believe that machine learning (ML) techniques can be used to reduce false alarms. We have thoroughly investigated several ML-based attack detection methods in SDN. There are two ML-based simulation approaches simulation data and a public dataset. We completed our experiment on all these studies and explained why certain standard machine learning–based methods fail. A system based on one of the most efficient machine learning methods will be proposed in future work, considering some models' weaknesses and aiming to develop an improved model that will outperform current approaches in terms of accuracy and efficiency.

References

1. Abdullaziz, O.I., Wang, L.: Mitigating DoS attacks against SDN controller using information hiding. In: 2019 IEEE Wireless Communications and Networking Conference (WCNC), pp. 1–6. https://doi.org/10.1109/WCNC.2019.8885764 (2019)
2. Abubakar, A., Pranggono, B.: Machine learning based intrusion detection system for software defined networks. In: 2017 Seventh International Conference on Emerging Security Technologies (EST), pp. 138–143. https://doi.org/10.1109/EST.2017.8090413 (2017)

3. Adamou Djergou, A., Maleh, Y., Mounir, S.: Machine learning techniques for intrusion detection in SDN: a survey. In: Maleh, Y., Alazab, M., Gherabi, N., Tawalbeh, L., Abd El-Latif, A.A. (eds.) Advances in Information, Communication and Cybersecurity, pp. 460–473. Springer (2022)
4. Agborubere, B., Sanchez-Velazquez, E.: OpenFlow communications and TLS security in software-defined networks. In 2017 IEEE International Conference on Internet of Things (IThings) and IEEE Green Computing and Communications (GreenCom) and IEEE Cyber, Physical and Social Computing (CPSCom) and IEEE Smart Data (SmartData), pp. 560–566. https://doi.org/10.1109/iThings-GreenCom-CPSCom-SmartData.2017.88 (2017)
5. Ahmed, M.E., Kim, H.: DDoS attack mitigation in Internet of Things using software defined networking. In: 2017 IEEE Third International Conference on Big Data Computing Service and Applications (BigDataService), pp. 271–276. https://doi.org/10.1109/BigDataService.2017.41 (2017)
6. Ahuja, N., Singal, G., Mukhopadhyay, D., Kumar, N.: Automated DDOS attack detection in software defined networking. J. Netw. Comput. Appl. **187**, 103108 (2021). https://doi.org/10.1016/J.JNCA.2021.103108
7. Alcorn, J.A., Chow, C.E.: A framework for large-scale modeling and simulation of attacks on an OpenFlow network. In: 2014 23rd International Conference on Computer Communication and Networks (ICCCN), pp. 1–6. https://doi.org/10.1109/ICCCN.2014.6911848 (2014)
8. Al-Haj, S., Tolone, W. J.: FlowTable pipeline misconfigurations in software defined networks. In: 2017 IEEE Conference on Computer Communications Workshops (INFOCOM WKSHPS), pp. 247–252. https://doi.org/10.1109/INFCOMW.2017.8116384 (2017)
9. Alparslan, O., Gunes, O., Hanay, Y.S., Arakawa, S., & Murata, M.: Improving resiliency against DDoS attacks by SDN and multipath orchestration of VNF services. In: 2017 IEEE International Symposium on Local and Metropolitan Area Networks (LANMAN), pp. 1–3. https://doi.org/10.1109/LANMAN.2017.7972158 (2017)
10. Banitalebi Dehkordi, A., Soltanaghaei, M., Boroujeni, F.Z.: The DDoS attacks detection through machine learning and statistical methods in SDN. J. Supercomput. **77**(3), 2383–2415 (2021). https://doi.org/10.1007/s11227-020-03323-w
11. Bauer, R., Dittebrandt, A., Zitterbart, M.: GCMI: a generic approach for SDN control message interception. In: 2019 IEEE Conference on Network Softwarization (NetSoft), pp. 360–368. https://doi.org/10.1109/NETSOFT.2019.8806661 (2019)
12. Brooks, M., Yang, B.: A man-in-the-middle attack against OpenDayLight SDN controller. In: Proceedings of the 4th Annual ACM Conference on Research in Information Technology, pp. 45–49. https://doi.org/10.1145/2808062.2808073 (2015)
13. Carvalho, R.N., Bordim, J.L., Alchieri, E.A.P.: Entropy-Based DoS Attack Identification in SDN. In: 2019 IEEE International Parallel and Distributed Processing Symposium Workshops (IPDPSW), pp. 627–634. https://doi.org/10.1109/IPDPSW.2019.00108 (2019)
14. Chang, S., Park, Y., Babu, B.B.A.: Fast IP hopping randomization to secure hop-by-hop access in SDN. IEEE Trans. Netw. Serv. Manag. **16**(1), 308–320 (2019). https://doi.org/10.1109/TNSM.2018.2889842
15. Chi, P.-W., Kuo, C.-T., Guo, J.-W., Lei, C.-L.: How to detect a compromised SDN switch. In: Proceedings of the 2015 1st IEEE Conference on Network Softwarization (NetSoft), pp. 1–6. https://doi.org/10.1109/NETSOFT.2015.7116184 (2015)
16. Chin, T., Mountrouidou, X., Li, X., Xiong, K.: Selective packet inspection to detect DoS flooding using software defined networking (SDN). In: 2015 IEEE 35th International Conference on Distributed Computing Systems Workshops, pp. 95–99. https://doi.org/10.1109/ICDCSW.2015.27 (2015)
17. Chouhan, R.K., Atulkar, M., Nagwani, N.K.: A framework to detect DDoS attack in Ryu controller based software defined networks using feature extraction and classification. Appl. Intell. (2022). https://doi.org/10.1007/s10489-022-03565-6
18. Cui, H., Chen, Z., Yu, L., Xie, K., Xia, Z.: Authentication mechanism for network applications in SDN environments. In: 2017 20th International Symposium on Wireless Personal Multimedia Communications (WPMC), pp. 1–5. https://doi.org/10.1109/WPMC.2017.8301788 (2017)

19. Cziva, R., Jouët, S., Stapleton, D., Tso, F.P., Pezaros, D.P.: SDN-based virtual machine management for cloud data centers. IEEE Trans. Netw. Serv. Manag. **13**(2), 212–225 (2016)
20. Dargahi, T., Caponi, A., Ambrosin, M., Bianchi, G., Conti, M.: A survey on the security of stateful SDN data planes. IEEE Commun. Surv. Tutor. (2017)
21. D'Orsaneo, J., Tummala, M., McEachen, J., Martin, B.: Analysis of traffic signals on an SDN for detection and classification of a man-in-the-middle attack. In: 2018 12th International Conference on Signal Processing and Communication Systems (ICSPCS), pp. 1–9. https://doi.org/10.1109/ICSPCS.2018.8631762 (2018)
22. Dotcenko, S., Vladyko, A., Letenko, I.: A fuzzy logic-based information security management for software-defined networks. In: 16th International Conference on Advanced Communication Technology, pp. 167–171. https://doi.org/10.1109/ICACT.2014.6778942 (2014)
23. Dridi, L., Zhani, M.F.: SDN-guard: DoS attacks mitigation in SDN networks. In: 2016 5th IEEE International Conference on Cloud Networking (Cloudnet), pp. 212–217. https://doi.org/10.1109/CloudNet.2016.9 (2016)
24. Elsayed, M.S., Le-Khac, N.-A., Jurcut, A.D.: InSDN: a novel SDN intrusion dataset. IEEE Access. **8**, 165263–165284 (2020). https://doi.org/10.1109/ACCESS.2020.3022633
25. Feghali, A., Kilany, R., Chamoun, M.: SDN security problems and solutions analysis. In: 2015 International Conference on Protocol Engineering (ICPE) and International Conference on New Technologies of Distributed Systems (NTDS), pp. 1–5. https://doi.org/10.1109/NOTERE.2015.7293514 (2015)
26. Foerster, K., Ludwig, A., Marcinkowski, J., Schmid, S.: Loop-free route updates for software-defined networks. IEEE/ACM Trans. Networking. **26**(1), 328–341 (2018). https://doi.org/10.1109/TNET.2017.2778426
27. Gao, S., Li, Z., Xiao, B., Wei, G.: Security threats in the data plane of software-defined networks. IEEE Netw. **32**(4), 108–113 (2018). https://doi.org/10.1109/MNET.2018.1700283
28. Goksel, N., Demirci, M.: DoS attack detection using packet statistics in SDN. In: 2019 International Symposium on Networks, Computers and Communications (ISNCC), pp. 1–6. https://doi.org/10.1109/ISNCC.2019.8909114 (2019)
29. Hadem, P., Saikia, D.K., Moulik, S.: An SDN-based intrusion detection system using SVM with selective logging for IP Traceback. Comput. Netw. **191**, 108015 (2021). https://doi.org/10.1016/J.COMNET.2021.108015
30. Isong, B., Molose, R.R.S., Abu-Mahfouz, A.M., Dladlu, N.: Comprehensive review of SDN controller placement strategies. IEEE Access. **8**, 170070–170092 (2020). https://doi.org/10.1109/ACCESS.2020.3023974
31. Jäger, B., Röpke, C., Adam, I., Holz, T.: Multi-layer access control for SDN-based telco clouds. In: Buchegger, S., Dam, M. (eds.) In Nordic Conference on Secure IT Systems, pp. 197–204. Springer (2015)
32. Khakimov, A., Elgendy, I.A., Muthanna, A., Mokrov, E., Samouylov, K., Maleh, Y., El-Latif, A.A.A.: Flexible architecture for deployment of edge computing applications. Simul. Model. Pract. Theory. **114**, 102402 (2022). https://doi.org/10.1016/J.SIMPAT.2021.102402
33. Klaedtke, F., Karame, G.O., Bifulco, R., Cui, H.: Access control for SDN controllers. In: Proceedings of the Third Workshop on Hot Topics in Software Defined Networking, pp. 219–220. https://doi.org/10.1145/2620728.2620773 (2014)
34. Klaedtke, F., Karame, G.O., Bifulco, R., Cui, H.: Towards an access control scheme for accessing flows in SDN. In: Proceedings of the 2015 1st IEEE Conference on Network Softwarization (NetSoft), pp. 1–6. https://doi.org/10.1109/NETSOFT.2015.7116185 (2015)
35. Kokila, R.T., Selvi, S.T., Govindarajan, K.: DDoS detection and analysis in SDN-based environment using support vector machine classifier. In: 2014 Sixth International Conference on Advanced Computing (ICoAC), pp. 205–210. https://doi.org/10.1109/ICoAC.2014.7229711 (2014)
36. Lantz, B., Heller, B., McKeown, N.: A network in a laptop: rapid prototyping for software-defined networks. In: Proceedings of the 9th ACM SIGCOMM Workshop on Hot Topics in Networks. https://doi.org/10.1145/1868447.1868466 (2010)
37. Lévai, T., Pelle, I., Németh, F., Gulyás, A.: EPOXIDE: a modular prototype for SDN troubleshooting. SIGCOMM Comput. Commun. Rev. **45**(4), 359–360 (2015). https://doi.org/10.1145/2829988.2790027

38. Li, C., Wu, Y., Yuan, X., Sun, Z., Wang, W., Li, X., Gong, L.: Detection and defense of DDoS attack–based on deep learning in OpenFlow-based SDN. Int. J. Commun. Syst. **31**(5), e3497 (2018). https://doi.org/10.1002/dac.3497
39. Maleh, Y., Qasmaoui, Y., el Gholami, K., Sadqi, Y., Mounir, S.: A comprehensive survey on SDN security: threats, mitigations, and future directions. J. Reliab. Intell. Environ. (2022). https://doi.org/10.1007/s40860-022-00171-8
40. Masoud, M. Z., Jaradat, Y., Jannoud, I.: On preventing ARP poisoning attack utilizing Software Defined Network (SDN) paradigm. In 2015 IEEE Jordan Conference on Applied Electrical Engineering and Computing Technologies (AEECT), pp. 1–5. https://doi.org/10.1109/AEECT.2015.7360549 (2015)
41. Matsumoto, S., Hitz, S., Perrig, A.: Fleet: defending SDNs from malicious administrators. In: Proceedings of the Third Workshop on Hot Topics in Software Defined Networking, pp. 103–108. https://doi.org/10.1145/2620728.2620750 (2014)
42. McKeown, N., Anderson, T., Balakrishnan, H., Parulkar, G., Peterson, L., Rexford, J., Shenker, S., Turner, J.: OpenFlow: enabling innovation in campus networks. ACM SIGCOMM Comput. Commun. Rev. **38**(2), 69–74 (2008)
43. Mekky, H., Hao, F., Mukherjee, S., Zhang, Z.-L., Lakshman, T.V.: Application-aware data plane processing in SDN. In: Proceedings of the Third Workshop on Hot Topics in Software Defined Networking, pp. 13–18. https://doi.org/10.1145/2620728.2620735 (2014)
44. Midha, S., Triptahi, K.: Extended TLS security and defensive algorithm in OpenFlow SDN. In: 2019 9th International Conference on Cloud Computing, Data Science & Engineering (Confluence), pp. 141–146. https://doi.org/10.1109/CONFLUENCE.2019.8776607 (2019)
45. Mihai-Gabriel, I., Victor-Valeriu, P.: Achieving DDoS resiliency in a software defined network by intelligent risk assessment based on neural networks and danger theory. In: 2014 IEEE 15th International Symposium on Computational Intelligence and Informatics (CINTI), pp. 319–324. https://doi.org/10.1109/CINTI.2014.7028696 (2014)
46. Myint Oo, M., Kamolphiwong, S., Kamolphiwong, T., Vasupongayya, S.: Advanced support vector machine- (ASVM-) based detection for distributed denial of service (DDoS) attack on software defined networking (SDN). J. Comput. Netw. Commun. **2019**, 8012568 (2019). https://doi.org/10.1155/2019/8012568
47. Namal, S., Ahmad, I., Gurtov, A., Ylianttila, M.: Enabling secure mobility with OpenFlow. In: 2013 IEEE SDN for Future Networks and Services (SDN4FNS), pp. 1–5. https://doi.org/10.1109/SDN4FNS.2013.6702540 (2013)
48. Nanda, S., Zafari, F., DeCusatis, C., Wedaa, E., Yang, B.: Predicting network attack patterns in SDN using machine learning approach. In: 2016 IEEE Conference on Network Function Virtualization and Software Defined Networks (NFV-SDN), 167–172. https://doi.org/10.1109/NFV-SDN.2016.7919493 (2016)
49. Nguyen, T., Yoo, M.: Attacks on host tracker in SDN controller: investigation and prevention. In: 2016 International Conference on Information and Communication Technology Convergence (ICTC), pp. 610–612. https://doi.org/10.1109/ICTC.2016.7763545 (2016)
50. Nife, F., & Kotulski, Z. (2018). *New SDN-Oriented Authentication and Access Control Mechanism BT – Computer Networks* (P. Gaj, M. Sawicki, G. Suchacka, A. Kwiecień, Eds.), pp. 74–88. Springer
51. Niyaz, Q., Sun, W., Javaid, A.Y.: *A Deep Learning Based DDoS Detection System in Software-Defined Networking (SDN)*. https://doi.org/10.4108/eai.28-12-2017.153515 (2017)
52. Oktian, Y.E., Lee, S.G., Lee, H.J., Lam, J.H.: Distributed SDN controller system: a survey on design choice. Comput. Netw. **121**, 100–111 (2017). https://doi.org/10.1016/j.comnet.2017.04.038
53. Padekar, H., Park, Y., Hu, H., Chang, S.-Y.: Enabling dynamic access control for controller applications in software-defined networks. In: Proceedings of the 21st ACM on Symposium on Access Control Models and Technologies, pp. 51–61. https://doi.org/10.1145/2914642.2914647 (2016)
54. Pan, H., Li, Z., Zhang, P., Salamatian, K., Xie, G.: Misconfiguration checking for SDN: data structure, theory and algorithms. In: 2020 IEEE 28th International Conference on Network Protocols (ICNP), pp. 1–11. https://doi.org/10.1109/ICNP49622.2020.9259353 (2020)

55. Phan, T.V, Gias, T.M.R., Islam, S.T., Huong, T.T., Thanh, N.H., Bauschert, T.: Q-MIND: defeating stealthy DoS attacks in SDN with a machine-learning based defense framework. In: 2019 IEEE Global Communications Conference (GLOBECOM), pp. 1–6. https://doi.org/10.1109/GLOBECOM38437.2019.9013585 (2019)
56. Phan, T.V., Park, M.: Efficient distributed denial-of-service attack defense in SDN-based cloud. IEEE Access. **7**, 18701–18714 (2019). https://doi.org/10.1109/ACCESS.2019.2896783
57. Qasmaoui, Y., Haqiq, A.: Solid-flow: a flow rules security mechanism for SDN. In: 2017 3rd International Conference of Cloud Computing Technologies and Applications (CloudTech), pp. 1–7. https://doi.org/10.1109/CloudTech.2017.8284734 (2017)
58. Qazi, Z.A., Lee, J., Jin, T., Bellala, G., Arndt, M., Noubir, G.: Application-awareness in SDN. SIGCOMM Comput. Commun. Rev. **43**(4), 487–488 (2013). https://doi.org/10.1145/2534169.2491700
59. Qi, C., Wu, J., Hu, H., Cheng, G., Liu, W., Ai, J., Yang, C.: An intensive security architecture with multi-controller for SDN. In: 2016 IEEE Conference on Computer Communications Workshops (INFOCOM WKSHPS), pp. 401–402. https://doi.org/10.1109/INFCOMW.2016.7562109 (2016)
60. Raikar, M.M., Meena, S.M., Mulla, M.M., Shetti, N.S., Karanandi, M.: Data traffic classification in software defined networks (SDN) using supervised-learning. Procedia Comput. Sci. **171**, 2750–2759 (2020). https://doi.org/10.1016/J.PROCS.2020.04.299
61. Ranjbar, A., Komu, M., Salmela, P., Aura, T.: An SDN-based approach to enhance the end-to-end security: SSL/TLS case study. In: NOMS 2016–2016 IEEE/IFIP Network Operations and Management Symposium, pp. 281–288. https://doi.org/10.1109/NOMS.2016.7502823 (2016)
62. Saâdaoui, A., Ben Youssef Ben Souayeh, N., Bouhoula, A.: Automated and optimized formal approach to verify SDN access-control misconfigurations. In: Gao, H., Yin, Y., Yang, X., Miao, H. (eds.) International Conference on Testbeds and Research Infrastructure, pp. 96–112. Springer (2019)
63. Sadqi, Y., Maleh, Y.: A systematic review and taxonomy of web applications threats. Inf. Secur. J. Glob. Perspect. **31**(1), 1–27 (2022). https://doi.org/10.1080/19393555.2020.1853855
64. Sahoo, K.S., Tripathy, B.K., Naik, K., Ramasubbareddy, S., Balusamy, B., Khari, M., Burgos, D.: An evolutionary SVM model for DDOS attack detection in software defined networks. IEEE Access. **8**, 132502–132513 (2020). https://doi.org/10.1109/ACCESS.2020.3009733
65. Schehlmann, L., Abt, S., Baier, H.: Blessing or curse? Revisiting security aspects of Software-Defined Networking. In: 10th International Conference on Network and Service Management (CNSM) and Workshop, pp. 382–387. https://doi.org/10.1109/CNSM.2014.7014199 (2014)
66. Schueller, Q., Basu, K., Younas, M., Patel, M., Ball, F.: A hierarchical intrusion detection system using support vector machine for SDN network in cloud data center. In: 2018 28th International Telecommunication Networks and Applications Conference (ITNAC), pp. 1–6. https://doi.org/10.1109/ATNAC.2018.8615255 (2018)
67. Sebbar, A., Boulmalf, M., Kettani, M.D.E.C.El., Baddi, Y.: Detection MITM attack in multi-SDN controller. In: 2018 IEEE 5th International Congress on Information Science and Technology (CiSt), pp. 583–587. https://doi.org/10.1109/CIST.2018.8596479 (2018)
68. Shaghaghi, A., Kaafar, M.A., Buyya, R., Jha, S.: Software-Defined Network (SDN) Data Plane Security: Issues, Solutions and Future Directions. ArXiv Preprint ArXiv:1804.00262 (2018)
69. Shuangyu, H., Jianwei, L., Jian, M., Jie, C.: Hierarchical solution for access control and authentication in software defined networks. In: Au, M.H., Carminati, B., Kuo, C.-C.J. (eds.) International Conference on Network and System Security, pp. 70–81. Springer (2014a)
70. Sudar, K.M., Deepalakshmi, P.: Comparative study on IDS using machine learning approaches for software defined networks. Int. J. Intell. Enterp. **7**(1–3), 15–27 (2020). https://doi.org/10.1504/IJIE.2020.104642
71. Tang, T.A., Mhamdi, L., McLernon, D., Zaidi, S.A.R., Ghogho, M.: Deep recurrent neural network for intrusion detection in SDN-based Networks. In: 2018 4th IEEE Conference on Network Softwarization and Workshops (NetSoft), pp. 202–206. https://doi.org/10.1109/NETSOFT.2018.8460090 (2018)

72. Wang, H.: Authentic and confidential policy distribution in software defined wireless network. In: 2014 International Wireless Communications and Mobile Computing Conference (IWCMC), pp. 1167–1171. https://doi.org/10.1109/IWCMC.2014.6906520 (2014)
73. Wang, M., Liu, J., Chen, J., Liu, X., Mao, J.: PERM-GUARD: authenticating the validity of flow rules in software defined networking. In: Proceedings – 2nd IEEE International Conference on Cyber Security and Cloud Computing, CSCloud 2015 – IEEE International Symposium of Smart Cloud, IEEE SSC 2015, 37, pp. 127–132. https://doi.org/10.1109/CSCloud.2015.89 (2016)
74. Yan, Z., Zhang, P., Vasilakos, A.V.: A security and trust framework for virtualized networks and software-defined networking. Secur. Commun. Netw. **9**(16), 3059–3069 (2016). https://doi.org/10.1002/sec.1243
75. Yue, M., Wang, H., Liu, L., Wu, Z.: Detecting DoS attacks based on multi-features in SDN. IEEE Access. **8**, 104688–104700 (2020). https://doi.org/10.1109/ACCESS.2020.2999668
76. Zhang, H., Cai, Z., Liu, Q., Xiao, Q., Li, Y., Cheang, C.F.: A survey on security-aware measurement in SDN. In: Security and Communication Networks, 2018 (2018)
77. Zhang, L., Wang, Z., Gu, K., Miao, F., Guo, Y.: Transparent synchronization based port mutation scheme in SDN network. In: 2016 5th International Conference on Computer Science and Network Technology (ICCSNT), pp. 581–585. https://doi.org/10.1109/ICCSNT.2016.8070225 (2016a)
78. Zhang, L., Wei, Q., Gu, K., Yuwen, H.: Path hopping based SDN network defense technology. In: 2016 12th International Conference on Natural Computation, Fuzzy Systems and Knowledge Discovery (ICNC-FSKD), pp. 2058–2063. https://doi.org/10.1109/FSKD.2016.7603498 (2016b)
79. Zhou, H., Wu, C., Yang, C., Wang, P., Yang, Q., Lu, Z., Cheng, Q.: SDN-RDCD: a real-time and reliable method for detecting compromised SDN devices. IEEE/ACM Trans. Networking. **26**(5), 2048–2061 (2018). https://doi.org/10.1109/TNET.2018.2859483

Machine Learning–Based Identity and Access Management for Cloud Security

Harun Jamil, Abid Ali, Meryem Ammi, Ruslan Kirichek, Mohammed Saleh Ali Muthanna, and Faisal Jamil

1 Introduction

Cloud computing is one of the primary sources to provide services to end users. It consists of virtual machines (VMs), services, resource storage, processing, and task handling. In coordination with the cloud, integrating artificial intelligence (AI) makes the services more reliable and practical from a services point of view [1]. Cloud services allow users easy access to their personal or private information from related databases and ensure the availability of services processed under linked

H. Jamil
Department of Electronic Engineering, Jeju National University, Jeju, South Korea

A. Ali (✉)
Department of Computer Science, University of Engineering and Technology Taxila, Taxila, Punjab, Pakistan

M. Ammi
Criminal Justice College, Naif Arab University for Security Sciences, Riyadh, Saudi Arabia
e-mail: mammi@nauss.edu.sa

R. Kirichek
Software Engineering and Computer Science Department, The Bonch-Bruevich Saint-Petersburg State University of Telecommunications, St. Petersburg, Russian Federation
e-mail: kirichek@sut.ru

M. S. A. Muthanna
Department of Mathematics & Mechanics, Saint Petersburg State University, St Petersburg, Russia

F. Jamil
Faculty of Information Technology and Electrical Engineering, Department of ICT and Natural Sciences, Norwegian University of Science and Technology (NTNU), Trondheim, Norway
e-mail: faisal.jamil@ntnu.no

A. A. Abd El-Latif et al. (eds.), *Secure Edge and Fog Computing Enabled AI for IoT and Smart Cities*, EAI/Springer Innovations in Communication and Computing,
https://doi.org/10.1007/978-3-031-51097-7_15

devices and privacy machines distributed over the Internet [2, 3]. Cloud services are reliable for the users, but security is required to keep the identity of the data and users. It is crucial to provide efficient services by maintaining the services available to the users and authenticating them while going through sensitive services subject to security and privacy [4]. As per security concerns, the cloud is a much more attractive site to target hackers. With AI, cloud security is improved with the ability to control and provide secure computational services [5]. It is difficult for users to create their identity each time they connect to some cloud service by filling up the online form and providing their personal and sensitive information. This practice generates a stream of personal and sensitive information that, at the same, possibly might be misused. Therefore, developing a suitable identity management system for cloud computing is crucial.

A significant requirement is that the users of cloud computing services must control which personal data is disclosed and in which manner this data is utilized to reduce the risk of fraud and identity theft: cloud computing frameworks and upgrading the cloud services to suit a more considerable number of clients. However, privacy protection and general security issues present a solid obstacle for cloud clients to adjust integrity-layer-based cloud computing frameworks [6]. As a service, information technology is utilized effectively for cloud processing over the network, providing cloud clients with incredible computing capacity with immense extra memory with minimal effort and cost. Privacy protection and security is the prime risk when approaching cloud computing. Cloud computing benefits Internet and network users through storage, processing, network, and similar task-handling benefits [7]. Cloud is an excellent source to enhance vehicular organizational data security through a secure platform. Dangers of distributed computing allude to criminal hacking, extreme access of maverick administrators, vulnerability concerning the whereabouts of information in a world, how and where thoughts of security-aware-cloud, element of protection change, and standards vary over the national limitations [8].

A few people also refer to the possibility of an online cyberwar or terrorism as an approach to cloud computing threats. The people and corporations are worried about how consistent privacy and security are kept up under this newly introduced environment. A VANET-like trust and reliability are the required factors by the user, while provisioning and processing an identity management system is an enhanced framework to ensure cloud security [9–11]. The purpose of the research is to propose a new AI-enabled security framework for cloud services. The machine learning–based security concept provides the actual identity for the clarification and enhancement of services providing the framework for the cloud identity system. The architecture presents the services that use the cloud and real-time computing, enhancing its contributions toward correctly identifying and modifying data. The selection is because cloud computing holds sensitive services distributed globally [12].

In the rest of the chapter, we organize the motivation for the work and state-of-the-art comparison of the proposed technique in the Introduction section. The system architecture is represented in the Architecture section of the chapter. In

Sect. 3, we provide transposition of authentication using federated identities. The platform for client authentication use of provisioning for cloud computing structure is defined in Sect. 4. In Sect. 5, we recommend future directions to promote cloud computing security.

1.1 Motivation

Clouds provide effective SaaS, IaaS, and PaaS to users and organizations. Besides all the services, the cloud provides the security that it offers its users, which is one of the essential aspects that enhances the cloud-based security and links the security policies with its user agreements. Privacy protection and security is the prime risk when approaching cloud computing. Cloud computing refers to the provision of hosted services to Internet users in a broad sense. The primary motivation for cloud security through organized identity is to secure cloud-based services. The essence of the authenticated users is fundamental. Another explanation is to use AI to enable the organized identity system for effective and efficient system processing [13].

1.2 Comparison

AI-enabled cloud security is one of the best suitable techniques to maintain cloud security through an organized identity system. This section compares the proposed technique with the current state-of-the-art research articles (Table 1).

2 Proposed System Architecture

The cloud computing framework can give interminable registering resources on a request basis due to the availability of high adaptability included in its nature, which also avoids the requirements of the cloud-specialist organizations for devices or resource registering. The accompanying engineering outline gives this [18]: We assure you that the privacy of customers' personal information is maintained using our newly designed architecture. We have made genuine safety efforts more than rapidly assigned provisions to secure the clients' information [19]. Generally, cloud computing consists of three major elements. The user(s) associated with the cloud for loading the data, the cloud services providers, and the user(s) accessing the cloud. The AI layer enhances cloud security through integrated features such as machine learning, neural networks, fuzzy logic, Natural Language Processing (NLP), and expert systems [20].

Figure 1 shows the complete system architecture. Initially, the users contacts with cloud services provider to get cloud services on their devices. The authentication

Table 1 Comparison with recent research

Control parameters	MAC approach [14]	ABAC approach [15]	DAC approach [16]	RBAC approach [17]	Proposed Methodology
Authentication failure	Environment dependent	Low	Very low	Job role dependent	Significantly less (ignorable)
Reusability	Not cleared	Multiple	Multiple	Multiple	Multiple platforms
User convenience	Varies from system to system	High level	High level	High level	High level
Cloud node overhead	Very low	Security level dependent	Very low	Very low	Very low
Performance	Security level deployment	High level	Security level dependent	Medium	High level
Failure at single point	Low	Not discussed	Authorized failure	Very low	Very low
Role assignments	Dependent on single node assignment	Not mentioned	Not discussed	Lower	Multiple

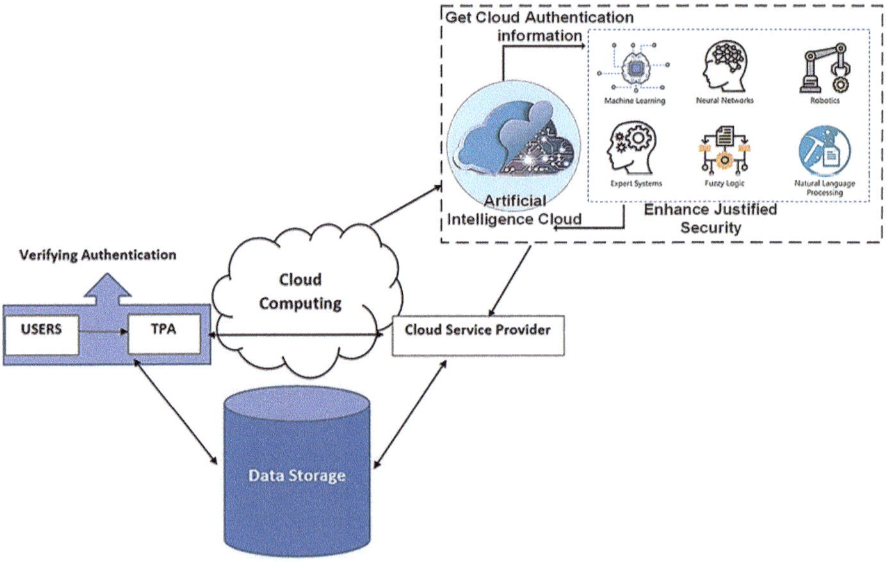

Fig. 1 Proposed system architecture for cloud services security using AI

of the users is fetched from data storage where old passwords are saved. The authentication information is passed to an AI-based module that checks AI-based authentication information and provides enhanced authentication security. The proposed system's components use clients, cloud services provider (CSP), multi-cloud database, and third-party auditor (TPA).

2.1 Clients

The client can be an individual group of individuals who have a place with an association. We have given a general classification of users, which incorporates all associated with stacking, getting to, and adjusting the information to the cloud. Such users are formally called the clients of the cloud [21]. Even if we can give sensible confirmations over the security and the trustworthiness of information put away in the cloud, a subsequent significant issue to consider is the security of client access to these resources to anticipate unapproved use. The customers can be the nonspecialized end user who connects with the admin's framework via some program through a passcode or graphical password [22, 23] or utilizing uses, just as a backup of storage disk or remote memory, the developer who uses a dynamic resource job in the cloud to boost up the apps or creating solutions or system's administrator who does not assemble the clouds yet sends onto them.

2.2 Cloud Service Provider (CSP)

As cloud services have been associated with the Internet, any issue associated with Internet security will also be a cause of influence for cloud computing. Services deployed in the cloud are obtainable through the Internet. Cloud specialist co-ops ensure the safety of their information and are mindful if any security threat affects their admin panel [24]. The services provided by a cloud provider have the potential to benefit their customers in a multitude of ways. These benefits include but are not limited to quick and convenient access to data from any location through a cloud-connected dashboard on various devices such as mobile phones or tablets, flexibility, pay-as-you-go pricing, secure data storage, retrieval, and protection against cyberattacks.

Additionally, customers can request security controls as needed and take advantage of the network and infrastructure facilities provided by the cloud supplier [25–27]. The CSP gets the requests and uses them to affirm the customer's profile from the concerned database. By adopting this procedure, we double-check the user's identity on the cloud. If the user identity is approved then data change on cloud. CSP's process in our framework is to authenticate the user's confirmation and keep up the cloud for its affiliation. The distributed storage supplier must have the option to give an encryption mapping to the put-away information, get to control for their information to keep an unapproved client from getting to the information, and give a reinforcement administration to their knowledge [28].

2.3 Multi-cloud Database

Repositioning cloud computing from a specific cloud for distributions to assure the privacy and security of users' provided information is considered critical. Figure 2 gives a general review of the cloud computing condition. Section A speaks to the customer side, which sends an information request to servers or examples in the cloud specialist organization to some degree. To some degree, the information source stores the information on the cloud side, which should be confided in the cloud. Moreover, the security of any question the customer has caused must be kept up. Yet, one cannot ensure the cloud gives confidence in the administration.

2.4 Third-Party Auditor (TPA)

Figure 2 shows the cloud user model. The proposed scheme's main layers are the user and cloud sides. The TPA utilizes a database for validating the client alone. When the client gets to the cloud just because their ID information, for example, the username and password, is stored in the specific database. The third-party auditor is used to verify whether the cloud gets an authenticated user. The user responds as username, password, and the requestor stacking gets to or changes the data in the cloud. TPA is responsible for supporting the user. TPA receives the information regarding the username and password of the client and stores it in the concerned database when the user is about to log in to the cloud. For every progressive login,

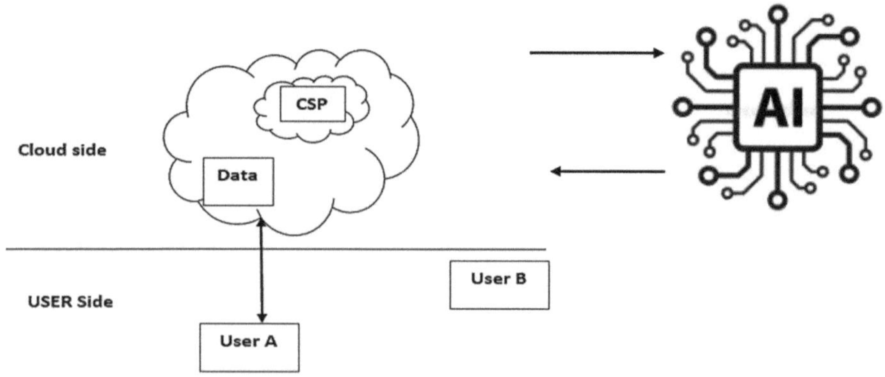

Fig. 2 General overview of user/cloud model

TPA performs verification of the user. Once the TPA validates the authenticity of the user, it sends the request to the cloud service provider.

3 Revolutionizing Authentication: A Deep Dive into Federated Identity Transposition

To ensure security, it is necessary to establish guidelines that outline the limits of user operations within a system, as well as the methods for managing, protecting, and distributing information and resources. This security arrangement should include a comprehensive set of rules governing the provision of system services. Having built up security strategies, frameworks must play out the accurate depictions of conduct to comply with the security models. These models are ordered in three fundamental sorts:

1. The discretionary model depends on identity or client data.
2. Mandatory or a necessary model which depends on rules.
3. The model depends on jobs or enables access to resources as indicated by an action performed on the framework.

The advances of web administrations center essentially around guaranteeing security in the collaborations between two obscure elements with distinctive security frameworks.

Most, by far, specialist co-ops are ensured by some verification systems. Along these lines, suppliers are required to track all clients. Consequently, clients should give their advanced character or convey enrollment in the framework to access the resources. Be that as it may, there are issues with a character on the board for clients and specialist co-ops since there is an incredible amount of data to deal with. The target of identity the board is to accomplish coordination of approaches, business procedures, and innovations to hold onto control of access to resources in a secure way, guaranteeing the classification of clients' data. The personalities can be overseen in three unique manners: conventional, centralized, and unified.

1. In the conventional model, each specialist co-op gives characters and certifications with the goal that their administrations can be gotten to. Every customer has unique and explicit identifiers for each assistance.
2. The centralized model depends on the rule that only a single identity supplier is answerable for overseeing characters that can be utilized in various divisions inside a similar space.
3. The unified model permits spaces where specialist organizations are just as suppliers of characters and accreditations. It is permitted that characters of a specific area get to different areas, given that previous understandings exist between the parts.

4 Platform for Client Authentication Use of Provisioning in Cloud Computing Structure

Figure 3 shows the confirmation of a client in a cloud computing condition and the related advances required for the equivalent. When a client registers, it includes individual data put away in a good area. A specialist co-op in a framework offers recognizable client proof (ID) to perform verification. The user sends a request by providing a username and password to the TPA. The sent request may be as follows:

1. *Request for putting data into the cloud.*
   The authenticated users may require sending or placing data into the cloud so that the remote users may interact with it as per their concerns. In this approach, the CSP may require a two-way user profile check before authenticating them to incorporate with the data.
2. *Request for getting the data in the cloud.*
   The associated remote users may require getting the data in the cloud. They utilize the web office to pick up induction to the CSP to request the data. The CSP responds to its request in the wake of checking its validity.
3. *Request for modifying the data in the cloud.*
   The authenticated users may require modifying or updating the data in the cloud. The CSP processes this request after the user verification. After authenticating the request, the TPA gives the users subtitles to the CSP. The CSP affirms the validity of the user by using their password. This is done by "mystery state checking." The CSP then uses its database, which contains the password created per authentication and TPA verification at first time verification. The system validated the password with the password obtained from the TPA in this process.

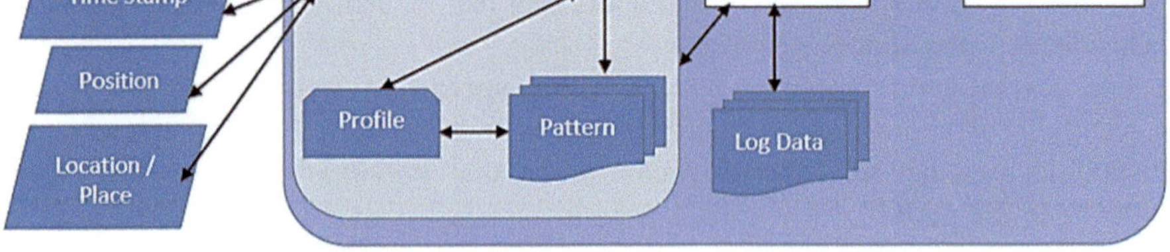

Fig. 3 Platform of approval

5 Results and Discussion

The simulation results are discussed and provide effective feedback for the proposed methodology. As this idea is used as AI and recommended identity system security for the cloud, we implement the results over the cipher cloud framework [29]. The simulation results described above are related to effective reasoning and enhance the working of the system effectiveness. The results effectively show in Fig. 3 that our proposed system effectively handles cloud-based authentication using AI techniques. The authenticated user system and monitoring agents provide adequate security based on the provided knowledge and data. After implementing the cloud-organized identity framework, the simulations are performed using Python. CloudSim simulation tool sets up the cloud VMs, servers, and target clients required to provide the services. The results are added based on the quality of services (QoS) parameter discussed in the proposed framework.

5.1 Dataset Description

The proposed content pre-caching scheme's effectiveness is demonstrated using numerical results from a real dataset trained and tested using supervised machine learning. We are developing a new approach to address the issue of cloud security using an organized identity system. The dataset is composed of self-generated data and data obtained from online sources. It is divided into training and testing sets, with a total of 115,585 records, and its key features are outlined with sample values.

5.2 Machine Learning Model Evaluation

To evaluate the effectiveness of our proposed system, we undergo training for each attribute class in the dataset. Subsequently, we test the system's performance using classification and model analysis tools for each class. The QoS results for every attribute are presented in Fig. 6, while the training accuracy of the model is depicted in Fig. 4. The training process improves the system's predictive abilities and offers valuable comparisons with other datasets.

$$\text{Accuracy} = \frac{\text{Number of Classified Samples}}{\text{Total Samples}} \tag{1}$$

From Fig. 5, it can be observed that the accuracy of the model rises swiftly as the predicted dataset range approaches its end. The model was trained by performing maximum iterations (epoch) on a dataset and monitoring the system's loss over the requested dataset. The system's performance was evaluated based on the loss of the

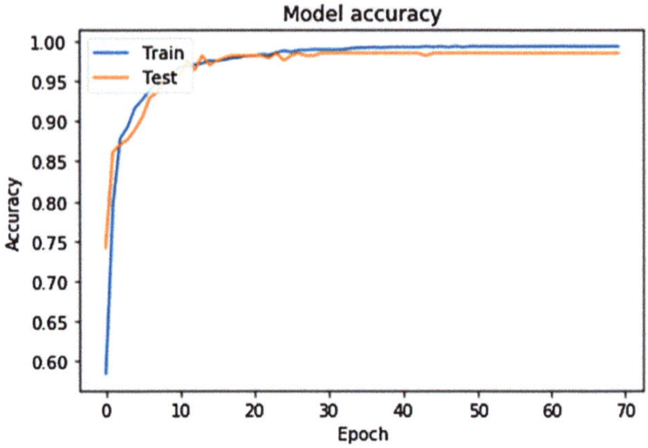

Fig. 4 Accuracy of the proposed machine learning CNN model

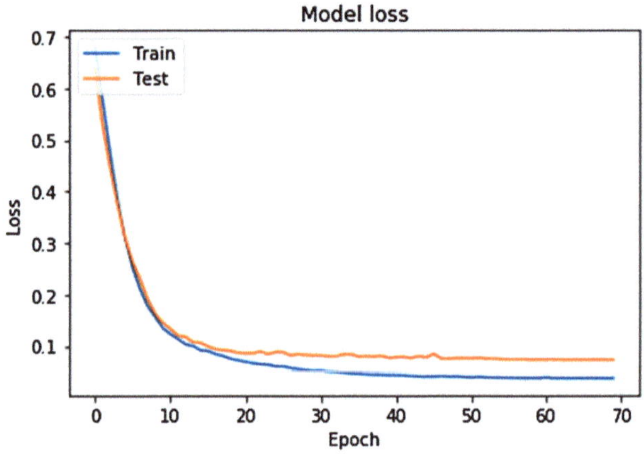

Fig. 5 Proposed CNN model loss with train over the testing dataset

model and measured using both the training and test datasets. Figure 6 illustrates that the proposed model has a shallow loss. The training and testing loss of the model was determined by utilizing a predefined dataset for the training and testing iterations of the proposed model.

5.3 Network Throughput

The network's VMs, infrastructure, and user processes work together cooperatively, resulting in successful content security and lower delays in cloud-based AI systems. The network's performance is evaluated based on its throughput, demonstrating reduced network congestion with the proposed technique. The protocol reduces the probability of data request collisions by implementing necessary features and enables the network to participate in overcoming congestion by utilizing all defined protocols. Figure 6 illustrates the network throughput of the proposed technique. In

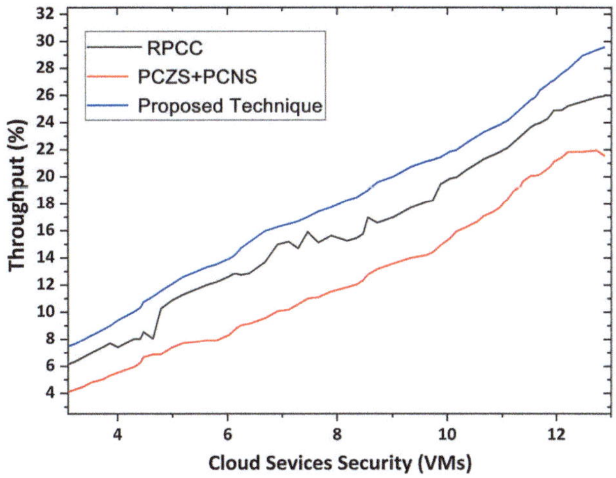

Fig. 6 Network throughput for congestion control in the proposed approach

a cloud security breach, a correlation exists between cloud services security (VMs) and the throughput.

6 Conclusions, Future Recommendations, and Research Implication

Even though cloud computing has expanded quickly, cloud computing security is a significant issue in the cloud computing environment. Clients prefer not to lose their private data because of malicious insiders inside the cloud. Additionally, the AI provides useful security features through its identified professional systems. Moreover, lost assistance accessibility has, as of late, caused numerous issues for countless cloud clients. The verification administrator checks the client's profile. Validation may depend on solid united substances. Integrated AI enhance the validation and verification of the client data. Data considered classified should be ensured. In this manner, suppliers look for mechanical developments that ensure the assurance and security of data clients. In this chapter, we investigated the issue of information security in distributed computing stockpiling, which is fundamentally an appropriate stockpiling system. To ensure the exactness of the customer's information in the distributed storage, we proposed a groundbreaking and versatile security framework in cloud processing with AI accessibility. Finally, AI enhances the cloud client's security for efficient data processing and accessibility issues.

6.1 Future Work

In the future, deep learning approaches will be used to secure the cloud authentication scheme. Furthermore, the enhanced security mechanism provides adequate security and extensibility for the cloud-based secure strategy.

6.2 Implications of the Study

The research is very relevant to the current state-of-the-art research problems identified in cloud services provision. The companies can adopt the research which provides cloud services to their client and wants to secure their services from external intruders using AI-enabled technology. Moreover, the research also applies to the current cloud-based infrastructure companies.

References

1. Ahmad, S., Khan, S., Jamil, F., Qayyum, F., Ali, A., Kim, D.: Design of a general complex problem-solving architecture based on task management and predictive optimization. Int. J. Distrib. Sensor Netw. **18**(6), 15501329221107868 (2022)
2. Razaque, A., Frej, M., Alotaibi, B., Alotaibi, M.: Privacy preservation Models for Third-Party Auditor over Cloud Computing: a Survey (2021)
3. Werner, J., Westphall, C.M., Westphall, C.B.J.C.N.: Cloud identity management: a survey on privacy strategies. Comput. Netw. **122**, 29–42 (2017)
4. Ali, A., et al.: An efficient dynamic-decision based task scheduler for task offloading optimization and energy management in mobile cloud computing. Sensors. **21**(13), 4527 (2021)
5. Ali, A., Iqbal, M.M.J.I.A.: A cost and energy efficient task scheduling technique to offload microservices based applications in mobile cloud computing. IEEE Access. **10**, 46633–46651 (2022)
6. Bowers, K.D., Juels, A., Oprea, A.: HAIL: a high-availability and integrity layer for cloud storage. In: Proceedings of the 16th ACM Conference on Computer and Communications Security, pp. 187–198 (2009)
7. Haji, L.M., Ahmad, O.M., Zeebaree, S., Dino, H.I., Zebari, R.R., Shukur, H.M.J.: Impact of cloud computing and internet of things on the future internet. Technol. Rep. Kansai Univ. **62**(5), 2179–2190 (2020)
8. Naidu, P.R., Guruprasad, N., Gowda, V.D.: A high-availability and integrity layer for cloud storage, cloud computing security: from single to multi-clouds. J. Phys. Conf. Ser. **1921**(1), 012072 (2021)
9. Torok, G., Day, M.R., Hartman-Baker, R.J., Snavely, C.: Iris: allocation banking and identity and access management for the exascale era. In: SC20: International Conference for High Performance Computing, Networking, Storage and Analysis, pp. 1–11. IEEE (2020)
10. Pöhn, D., Hommel, W.: Proven and modern approaches to identity management. In: Advances in Cybersecurity Management, pp. 421–443. Springer (2021)
11. Diaz-Piraquive, F.N., Ortiz-Ruiz, E., González-Aragón, J.J., Avila-Bermeo, H.H., Parada-Jaimes, W.D.: Knowledge management applied in the comparative study of the IRETE

intrusion methodology to access to the computer systems. In: International Conference on Knowledge Management in Organizations, pp. 354–365. Springer (2021)
12. Ali, A., Jamil, F., Whangbo, T., Ahmad, S.: AI-Enabled Cybernetic Analytics of Security Models for Smart Serious Games-Based Mobile Operating Systems
13. Ali, A., et al.: VABLOCK: a blockchain-based secure communication in V2V network using icn network support technology. Microprocess. Microsyst. **93**, 104569 (2022)
14. Zhang, Z., et al.: Artificial intelligence in cyber security: research advances, challenges, and opportunities. Artif. Intell. Rev. **55**(2), 1029–1053 (2022)
15. Singh, P., Singh, N., Luxmi, P.R., Saxena, A.: Artificial intelligence for smart data storage in cloud-based IoT. In: Transforming Management with AI, Big-Data, and IoT, pp. 1–15. Springer (2022)
16. Ahmad, S., Mehfuz, S., Beg, J.: Empirical analysis of security enabled quantum computing for cloud environment. In: Quantum and Blockchain for Modern Computing Systems: Vision and Advancements, pp. 103–125. Springer (2022)
17. N. J. S. C. Moustafa and Society: A new distributed architecture for evaluating AI-based security systems at the edge: network TON_IoT datasets. Sustain. Cities Soc. **72**, 102994 (2021)
18. Abd El-Latif, A.A., Abd-El-Atty, B., Hossain, M.S., Elmougy, S., Ghoneim, A.J.I.A.: Secure quantum steganography protocol for fog cloud internet of things. IEEE Access. **6**, 10332–10340 (2018)
19. Tabrizchi, H., Rafsanjani, M.K.: A survey on security challenges in cloud computing: issues, threats, and solutions. J. Supercomput. **76**(12), 9493–9532 (2020)
20. Li, Y., Kumar, R., Lasecki, W.S., Hilliges, O.: Artificial intelligence for HCI: A modern approach. In: Extended Abstracts of the 2020 CHI Conference on Human Factors in Computing Systems, pp. 1–8 (2020)
21. Abd-El-Atty, B., Iliyasu, A.M., Alaskar, H., El-Latif, A., Ahmed, A.J.S.: A robust quasi-quantum walks-based steganography protocol for secure transmission of images on cloud-based E-healthcare platforms. Sensors. **20**(11), 3108 (2020)
22. Andrew, S., Watson, S., Oh, T., Tigwell, G.W.: A review of literature on accessibility and authentication techniques. In: The 22nd International ACM SIGACCESS Conference on Computers and Accessibility, pp. 1–4 (2020)
23. Jagadamba, G., Sheeba, R., Brinda, K., Rohini, K., Pratik, S.: Adaptive e-learning authentication and monitoring. In: 2020 2nd International Conference on Innovative Mechanisms for Industry Applications (ICIMIA), pp. 277–283. IEEE (2020)
24. Rizvi, S., Mitchell, J., Razaque, A., Rizvi, M.R., Williams, I.J.: A fuzzy inference system (FIS) to evaluate the security readiness of cloud service providers. J. Cloud Comput. **9**(1), 1–17 (2020)
25. Alakbarov, R.G.: Challenges of mobile devices' resources and in communication channels and their solutions. Int. J. Comp. Netw. Inf. Secur. **13**(1), 39–46 (2021)
26. Fellah, H., Mezioud, C., Batouche, M.C.: Mobile cloud computing: architecture, advantages and security issues. In: Proceedings of the 3rd International Conference on Networking, Information Systems & Security, pp. 1–9 (2020)
27. Butt, U.A., Amin, R., Mehmood, M., Aldabbas, H., Alharbi, M.T., Albaqami, N.: Cloud security threats and solutions: A survey. Wirel. Pers. Commun. **128**(1), 387–413 (2023)
28. Abd El-Latif, A.A., et al.: Secret images transfer in cloud system based on investigating quantum walks in steganography approaches. Phys. A Stat. Mech. Appl. **541**, 123687 (2020)
29. Alibrahemi, D.A.M.: Block RADG Cipher Automata in Cloud Computing Networks. Ministry of Higher Education (2018)

Spatial Data of Smart Cities: Trust

Alexandr Shestakov and Alexey Nesterov

1 Spatial Data: Smart City

Different institutions and organizations involved in the development of the infrastructure of the smart city community determine the essence, requirements, nomenclature, and quality of spatial data. At the same time, these institutions implement their own understanding of spatial data and approaches to assessing trust in them, depending on the goals and objectives that these organizations solve.

The smart city concept from the perspective of the International Organization for Standardization [5–11], the International Telecommunication Union [12], the International Electrotechnical Commission, and the European Institute for Telecommunication Standardization [13, 14] is a mechanism for effective integration of information and communication technologies, including the Internet of Things and the industrial Internet, artificial intelligence, real-time data analysis, information modeling of buildings, and unmanned transport, in order to improve the standard of population living, economic development and management of urban services, and infrastructure and services provided [3, 4].

The existing dynamics of publication and inventive activity on the problems of smart cities is presented in Fig. 1 [2].

The results of the analysis of the electronic spatial data resources of smart cities show that in August 2022, for example, the number of spatial data sites visits, located in Moscow (Russia), was rgis.mosreg.ru – 145,8K; isogd.mosreg.ru –

A. Shestakov (✉) · A. Nesterov
Bonch-Bruevich Saint-Petersburg State University of Telecommunications, Saint-Petersburg, Russia
e-mail: nesterov.aa@spbgut.ru

A. A. Abd El-Latif et al. (eds.), *Secure Edge and Fog Computing Enabled AI for IoT and Smart Cities*, EAI/Springer Innovations in Communication and Computing,
https://doi.org/10.1007/978-3-031-51097-7_16

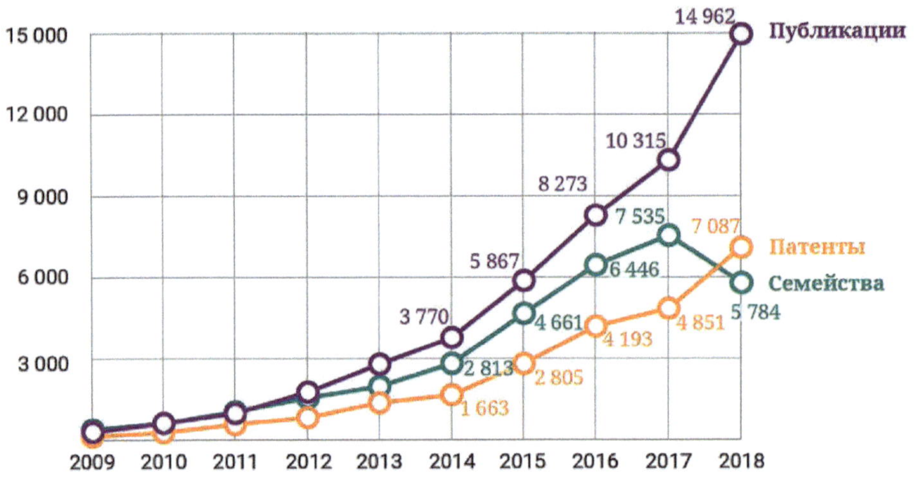

Fig. 1 Growth rates of "smart cities" research. (Source: [2])

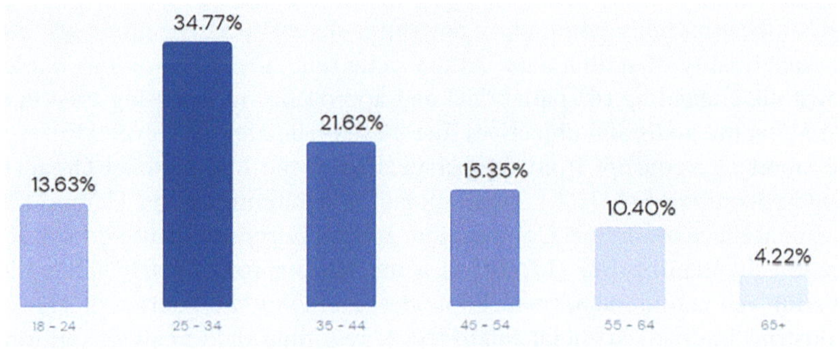

Fig. 2 The age distribution of audience interest in isogd.mosreg.ru

132,5K.[1] The age distribution of audience interest in spatial data sites is almost the same (Figs. 2 and 3).

The situation needs to clarify the common understanding of spatial data in cyber-physical systems for consistency, time comparability, or comparability in different systems to standardized data exchange and/or sharing and effective mechanisms for information exchange between organizations (institutions) in the field of responsibility for smart city infrastructures [15–19].

The most correct, in our opinion, is the following definition of spatial data of smart cities.

Spatial data of the smart city – data about spatial objects (point, linear, areal, and so on) of the territories, water areas, airspace, or the bowels of the Earth of natural or man-made origin in cyber-physical systems and an identifiable set of such data (sets of spatial data) that do not contradict the concepts of ISO 19115 [20], ISO 19131

[1] https://www.similarweb.com/ru/website/isogd.mosreg.ru/#overview

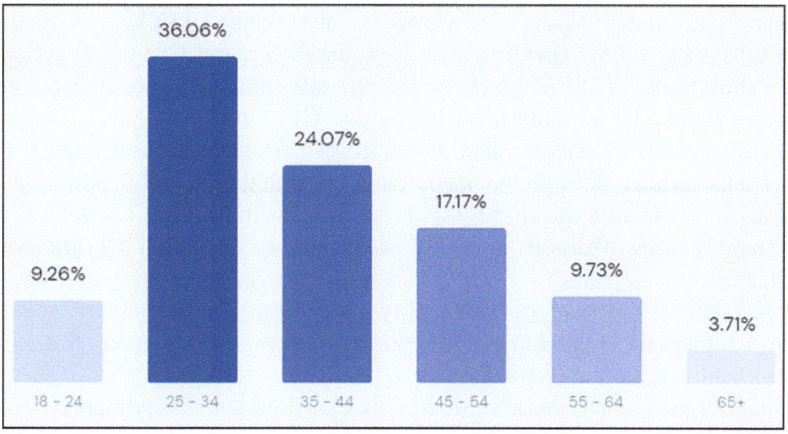

Fig. 3 The age distribution of audience interest in rgis.mosreg.ru

[21], having the ability to quickly determine and register, create, receive, collect, process, and transform them in understanding the quality of data according to ISO 19157 [22], requirements and estimates – according to ISO/IEC 250mn (Systems and Software Quality Requirements and Evaluation, SQuaRE).

In the definition, everything that is indicated before the formulation of the purpose of spatial data refers to one of the types of provision of a smart city and its infrastructures – to geoinformation support, that is, to the technological essence of a smart city. The user's interest lies precisely in the demand and the possibility of using this data for the target tasks of a smart city, as well as in the expected trust in such specific data. In particular, the accessibility and the possibility of guaranteed receipt of the required spatial data and metadata, for example, by an unqualified user in online access mode, should be ensured, and a simple, understandable tool that is trusted should be used.

2 Spatial Data: Confidence Assessment

Indicators of trust in spatial data of smart cities should be in a certain correlation relationship with those indicators of a smart city that operate with the categories "spatial object" of a smart city within its infrastructures in accordance with accepted models for assessing the development of a smart city.

The results of the conducted research on the systematization of models for assessing the development of smart cities, for example, in [1, 3, 23, 24], indicate the possibility of grouping them as follows:

- *Level's assessment models*: SCM (Smart City Maturity Model) 3–6 levels; SCR (Smart City Reference) 7 levels.
- *Component's assessment models*: ESCR (European Smart Cities Ranking) 6 directions, 62 indicators; Lorena Batagan, 7 directions, 49 indicators; SCIMI

(Smart City Index Master Indicators) 6 directions; CITYkeys 5 directions, 92 indicators; ENSC (the Ericsson Networked Society City) 8 directions, 28 indicators; CoO (City of Opportunity) 10 directions, 67 indicators; Cities in Motion Index, 10 directions, 83 indicators; GPCI (Global Power City Index), 6 directions, 70 indicators; data-driven cities (PricewaterhouseCoopers, PwC); smart city indicators (joint stock company "National Research Institute of Technologies and Communications") 7 directions, 26 indicators; Digital life Index of Russian cities (Moscow School of Management "Skolkovo") 7 directions, 18 indicators.

- *Profile assessment models*: Smart City Index (EasyPark company, specialization – transport), 4 directions, 19 indicators; eco-city dynamics, 5 directions; smart healthcare organizations.
- *Combined assessment models*: Smart City Development Assessment Matrix.

Formally The value of the Smart City Index (I), depending on the model, can be determined by the following expression:

$$I = \begin{cases} L_l, Q_{SD}, l = \overline{1, N}, \\ C_c, Q_{SD}, c = \overline{1, N}, \\ P_p, Q_{SD}, p = \overline{1, N}, \end{cases} \quad (1)$$

based on the results of the assessment of the values of the development indicators of the level (stage of development) (Ll), based on the results of the assessment of the values of the component indicators (Cc), or based on the results of the assessment of the values of the profile indicators (Pp), provided that the parameters of the characteristics of the essential properties of the spatial data of the smart city (QSD) are objectively present in the hierarchical system of indicators and in level by level, and component by component, and profile models.

The values of the parameters of the characteristics of the essential properties of the spatial data of a smart city in relation to the objectives of the development of a smart city can be interpreted as quality, and in relation to the procedures for the functioning and development of smart city infrastructures – as trust.

Lemma 1 The parameters of the characteristics of the essential properties of the spatial data of a smart city are included in at least one indicator of the level (stage of development) (L_l), or component (C_c), or profile (P_p).

Assessment of the quality of spatial data of a smart city is a set of procedures (operations), which includes the selection of a nomenclature of indicators of characteristics of essential properties, determining the values of these indicators, comparing them with prototypes and normalization, performing aggregation (convolution) taking into account the importance of the components in the final indicator in the following sequence: <private quality indicator> – <quality indicator> – <complex quality indicator> – <integral quality indicator> – <generalized quality indicator> [25].

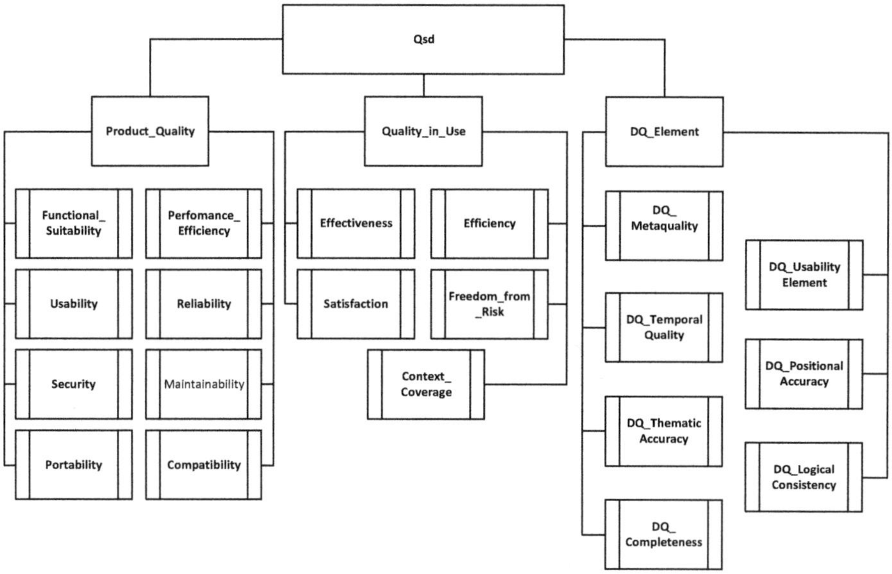

Fig. 4 The spatial data quality model

The basic nomenclature of indicators of the characteristics of the essential properties of the spatial data of the smart city can be represented as a set of indicators of the characteristics of the spatial data quality model (ISO 19157), the quality model, and the quality model when using SQuaRE (ISO/IEC 250mn) (Fig. 4).

To ensure the uniformity of the values of indicators having different units of measurement and to translate the values of such indicators into a single scale in the range [0, 1], a rationing procedure is applied. The calculation is carried out according to the formula:

$$Q_{SDi} = \frac{X_i - X_{\min}}{X_{\max} - X_{\min}}, \tag{2}$$

where Q_{SDi} – normalized values of the indicator X_i; X_i – the actual value of the indicator; $X_{\min}$ and $X_{\max}$ – the minimum and maximum values of the indicator of the characteristic of an essential property of spatial data.

3 Spatial Data: Ensuring Trust

Ensuring the trust of users in the spatial data of the smart city should be considered as one of the main tasks of organizational and technical measures in concepts, plans, and activities for the formation of the smart city infrastructure.

There are various measures to ensure the trust of spatial data, for example, maintaining the required level of data stability using data cybersecurity monitoring and management mechanisms [26].

The main *confidence measures* of spatial data of smart cities are the following:

- *Availability*: Creation and application of the spatial data infrastructure of the smart city, considering accessibility to existing national spatial data funds, for example, NSDI (National Spatial Data Infrastructure), CGDI (Canadian Geospatial Data Infrastructure), GDI-DE (Geospatial Data Infrastructure (Germany)), INSPIRE (Infrastructure for Spatial Information in Europe), and others.
- *Flexibility*: The use of a regulated limited number of spatial data exchange formats in smart city infrastructures from the existing nomenclature, for example, WMS, TMS, WMTS, WEF, DXF, GML, F1M, SFX, and others.
- *Confidence*: Regulation of procedures for determining a trusted authorized operator of the spatial data infrastructure of the smart city and monitoring its activities, as, for example, implemented in Moscow and St. Petersburg (Russia);
- *Adequacy*: Improving the quality of spatial data of the smart city using innovative methods and technologies for the rapid identification and registration, creation, reception, collection, processing and transformation of data on spatial objects, for example, using orthophotoplanes, geodetic measurement results, and laser scanning.
- *Relevance*: Proactive updating of spatial data of the smart city through procedures for updating data of the classification system and coding of information of the spatial data infrastructure of the smart city, analysis, and evaluation of the quality of updated basic spatial data when interacting with information systems – sources of spatial data, as shown in Fig. 5.

The systems should have only trusted sources or procedures for sifting and analyzing the exchange files of spatial data with an assessment of the relevance and quality of spatial data.

The evaluation results of the proposed mechanism for ensuring the trust of spatial data according to the indicators of the quality model (ISO/IEC 250mn, SQuaRE) show an increase in the values of indicators compared to systems without trust assurance procedures.

The main direction of further research on the mechanisms of ensuring the spatial data trust of the smart cities should be considered a more detailed analysis of the organization of spatial data of various types in the interests of solving user tasks of the smart city, methods of processing and spatial analysis of spatial data in the flow of heterogeneous data of complex cyber-physical systems of the smart city, and mechanisms of purposeful proactive impact on spatial data, including issues of analysis, modeling, optimization, and improvement of management and decision-making, in order to increase the efficiency of the functioning of the geoinformation infrastructure of the smart cities.

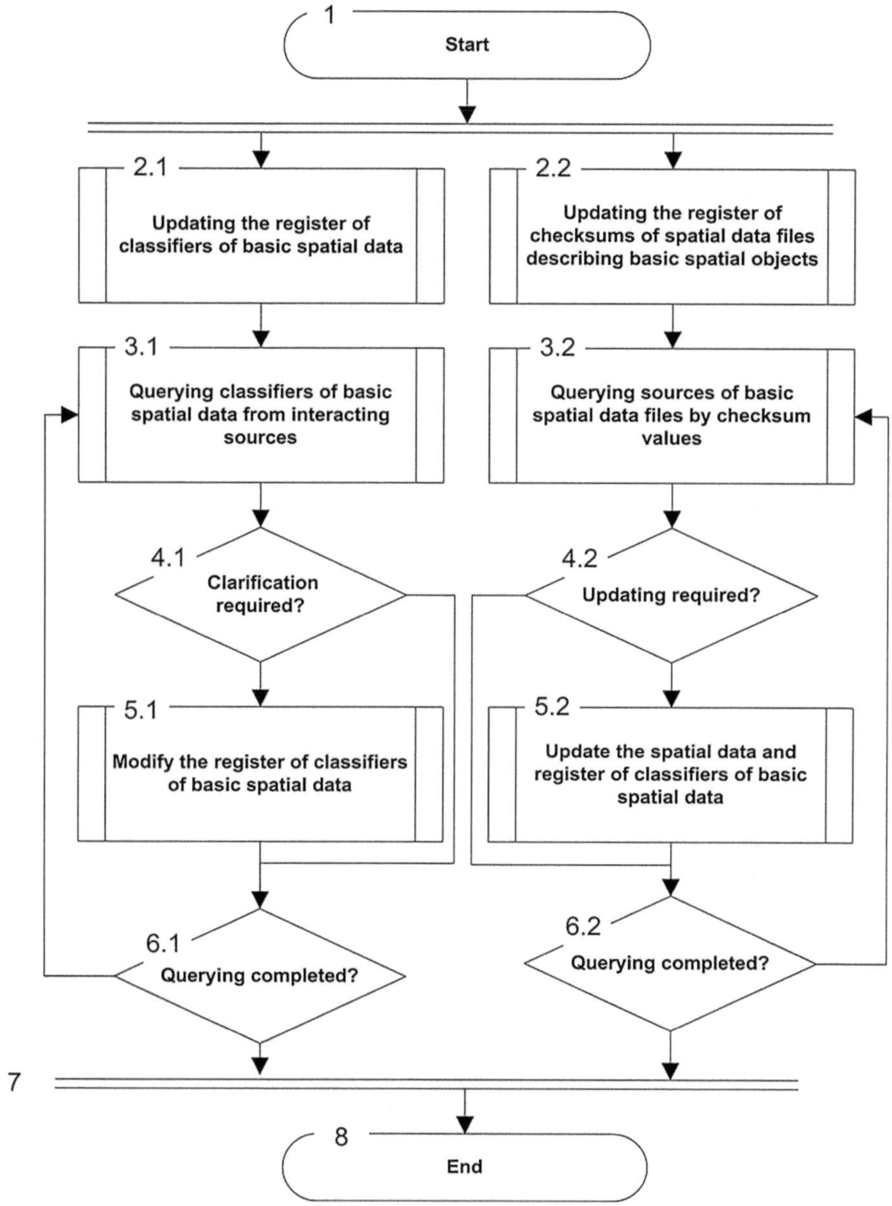

Fig. 5 Trust ensuring mechanism of spatial data

Acknowledgments The authors express their gratitude to the editors and reviewers for their attentive attitude to the chapter and the above comments, which made it possible to improve its quality. The authors declare that there is no conflict of interest. The chapter was prepared within the framework of applied scientific research conducted by the Bonch-Bruevich Saint-Petersburg State University of Telecommunications in 2022.

References

1. Prioritetnyye napravleniya vnedreniya tekhnologiy umnogo goroda v rossiyskikh gorodakh. Ekspertno-analiticheskiy doklad. M.: Tsentr strategicheskikh razrabotok «Severo-Zapad».175 s (2018)
2. Patentnyy landshaft «Umnyy gorod» -M.: Proyektnyy ofis FIPS, 168 s. (2019)
3. Popov YE, V., Semyachkov, K.A.: Matritsa pokazateley razvitiya umnykh gorodov. Natsional'nyye interesy: prioritety i bezopasnost, T. 16, vyp. 8. -S. 1422–1443 (2020)
4. Tsifrovizatsiya rossiyskikh gorodov: reyting otkrytykh sistem prostranstvennykh dannykh. M.: Fond "Institut ekonomiki goroda", 29 s. (2020)
5. ISO 37101:2016 Sustainable Development in Communities – Management System for Sustainable Development – Requirements with Guidance for Use
6. ISO 37120:2018 – Sustainable Development of Communities – Indicators for City Services and Quality of Life
7. ISO 37122:2019 Sustainable Cities and Communities – Indicators for Smart Cites
8. ISO 37123:2019 Sustainable Cities and Communities – Indicators for Resilient Cities
9. ISO/TR 37152:2016, "Smart community infrastructures – Common framework for development and operation", MOD
10. ISO 37155-1:2020 Information Technology. Smart City. Framework for Integration and Operation of Smart City Infrastructures – Part 1: Recommendations for Considering Opportunities and Challenges from Interactions in Smart Community Infrastructures from Relevant Aspects through the Life Cycle, MOD
11. ISO 37156:2020 Smart community infrastructures – Guidelines on data exchange and sharing for smart community infrastructures/
12. The Manizales Manifesto "Circular Economy and Smart Sustainable Cities & Communities" Adopted by the participants of the 7th Green Standards Week in Manizales, Colombia, on 5 April 2017
13. Pokazateli «umnykh» ustoychivykh gorodov, razrabotannyye YEEK OON-MSE. Zapiska sekretariata. Sem'desyat shestaya sessiya (Zheneva, 14–15 dekabrya 2015 goda). – Zheneva: Komitet po zhilishchnomu khozyaystvu i zemlepol'zovaniyu YEEK OON. 14 s. https://www.unece.org/fileadmin/DAM/hlm/documents/2015/ECE_HBP_2015_4.ru.pdf (2015)
14. Huovilaa, A., Boschb, P., Airaksinen, M.: Comparative analysis of standardized indicators for smart sustainable cities: what indicators and standards to use and when? Cities. **89**, 141–153 (2019) www.elsevier.com/locate/cities
15. Bhattacharya, D., Painho, M.: Design for Connecting Spatial Data Infrastructures with Sensor Web (SENSDI). ISPRS Annals of the Photogrammetry, Remote Sensing and Spatial Information Sciences, Volume III-4, XXIII ISPRS Congress, 12–19 July 2016, Prague, Czech Republic (2016)
16. Bhattacharya, D., Painho, M.: Smart Cities Intelligence System (SMACiSYS) Integrating Sensor Web with Spatial Data Interstructures (SENSDI). ISPRS annals of the photogrammetry, remote sensing and spatial information sciences, Volume IV-4/W3. In: 2017 2nd International Conference on Smart Data and Smart Cities, 4–6 October 2017, Puebla, Mexico (2017)
17. Santhanavanich, T., Schneider, S., Rodrigues, P., Coors, V.: Integration and visualization of heterogeneous sensor data and geospatial information. ISPRS annals of the photogrammetry, remote sensing and spatial information sciences, Volume IV-4/W7. In: 3rd International Conference on Smart Data and Smart Cities, 4–5 October 2018, Delft, The Netherlands (2018)
18. Chaturvedi, K., Matheus, A., Nguyen, S.H., Kolbe, T.H.: Securing spatial data infrastructures for distributed Smart City applications and services. Futur. Gener. Comput. Syst. **101**, 723–736 (2019)
19. Shestakov, A.V., Pletnev, Y.A., Frolova, K.A.: "Geoinfomationnyye sistemy v upravlenii i monitoring tekhnogennykh ob" yektov». Skhemy i QR-ssylki in, Sankt-Peterburg: SPb.: Tipografiya Lyubavich, p. 100 (2021)
20. ISO 19115:2014 Geographic information – Metadata

21. ISO 19131:2007 Geographic information – Data product specifications
22. ISO 19157:2013 "Geographic information – Data quality"
23. Popov Ye, V., Semyachkov, K.A: Sistematizatsiya podkhodov k otsenke razvitiya Umnykh gorodov. Ekonomika regiona (2020). T.16., vyp. 1. S. 14–27. https://doi.org/10.17059/2020-1-2
24. Indikatory umnykh gorodov NIITS – 2017. NIITS [Elektronnyy resurs]. 2017. Rezhim dostupa: http://niitc.ru/publications/SmartCities.pdf
25. Shestakov, A.V.: Vvedenie v metodologiu obrabotki geoprostranstvennuh dannuh genotipa telekommunikasei. Sankt-Peterburg:SPb.: GUAP, p. 325 (2016)
26. Solov'yev, A.V.: Algoritm otsenki ustoychivosti tsifrovykh dannykh. Trudy ISA RAN. Tom 71. 3/2021. S.47–52. https://doi.org/10.14357/20790279210305

Smart City Infrastructure Projects: Spatial Data of Risks

Kristina Frolova

1 Infrastructure Projects in Smart City

A smart city is the concept of integrating multiple information and communication technologies to manage urban infrastructures, such as healthcare, transport, environment, medicine, public safety, public services, urban management, communications and others. Each of them can be represented as an infrastructure project with its own phasing. The project contains different types of elements (subsystems) with heterogeneous links, which should be characterized by expansion (development or improvement). It is also worth noting the force majeure circumstances in which a project can degrade.

The problem of urban infrastructure management is exacerbated when there is an increasing degree of interconnectedness between different projects, as well as an increasing number of risks arising from their implementation.

For instance, in Moscow (Russia) according to the Unified Resources and Systems Register of Moscow [1] as of 20.09.2022, 170 information systems are registered and in use, which provide consolidated coordination and management of all smart city infrastructures.

Smart city infrastructure projects encompass all areas and, in their development, should ensure an improved quality of life. An example of a smart city structure is shown in Fig. 1.

The functional composition of infrastructure projects may include:

- *Housing and utilities*: Smart meters, leakage control, emergency management, consumption control [2].

K. Frolova (✉)
The Bonch-Bruevich Saint-Petersburg State University of Telecommunications, Saint Petersburg, Russia

A. A. Abd El-Latif et al. (eds.), *Secure Edge and Fog Computing Enabled AI for IoT and Smart Cities*, EAI/Springer Innovations in Communication and Computing,
https://doi.org/10.1007/978-3-031-51097-7_17

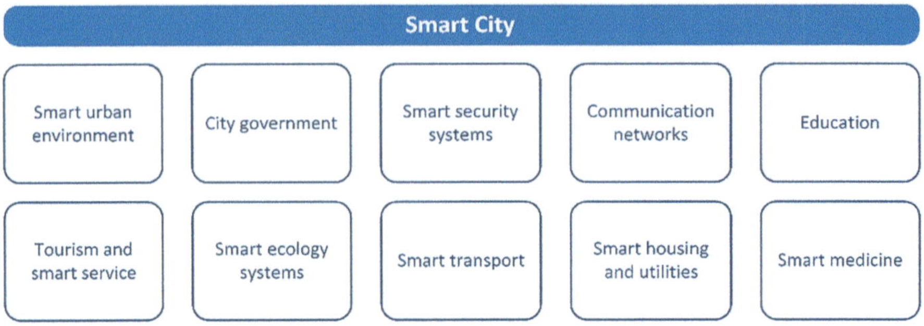

Fig. 1 Types of infrastructure projects in smart city

- *Smart transport*: Intelligent transport systems, smart parking [3, 4].
- *Smart urban environment*: Smart lighting, video surveillance, smart public safety systems, urban management, smart waste management, social services [5, 6].
- *Smart home*: Remote control of buildings, flats, integrated automation, smart home appliances, smart applications and services [7].

The data sources in the smart city infrastructure can be various sensors, video cameras, information systems and other infrastructures.

When implementing smart city infrastructures, a number of challenges can arise (Fig. 2). One of them is the inconsistency of data from different infrastructures. For example, city infrastructure data are managed by governmental organisations that may limit their use. The data-sharing approach will improve the ability to provide services in the smart city [8]. Another problem is the influence of city infrastructures on each other, as well as the influence of external sources. The solution to such a problem can be specifications that take into account mutual influence, as well as planning taking into account infrastructure risks [9].

The interaction of different infrastructures in the smart city includes:

- Information exchange between infrastructures.
- The risks of the impact of one infrastructure outcome on another.

The data framework for smart city and city infrastructure classifies data as metadata, reference or thematic data [8]. It is important to note among reference data the category of spatial data. Due to the geographical location of infrastructure objects, one of their attributes is spatial data.

Spatial data of the smart city are geographical data of urban infrastructure objects used for spatial analysis. Geographic information systems (GIS) have become an integral part of the smart city. GIS is not a new solution, but it has new uses all the time.

For example, all Polish cities participating in the study [10] use GIS tools, and 80% of them have implemented smart city tools. However, automatic integration of GIS and smart city tools is implemented only in 2 out of 15 surveyed cities. The function of overlaying thematic layers of GIS and smart city exists only in one

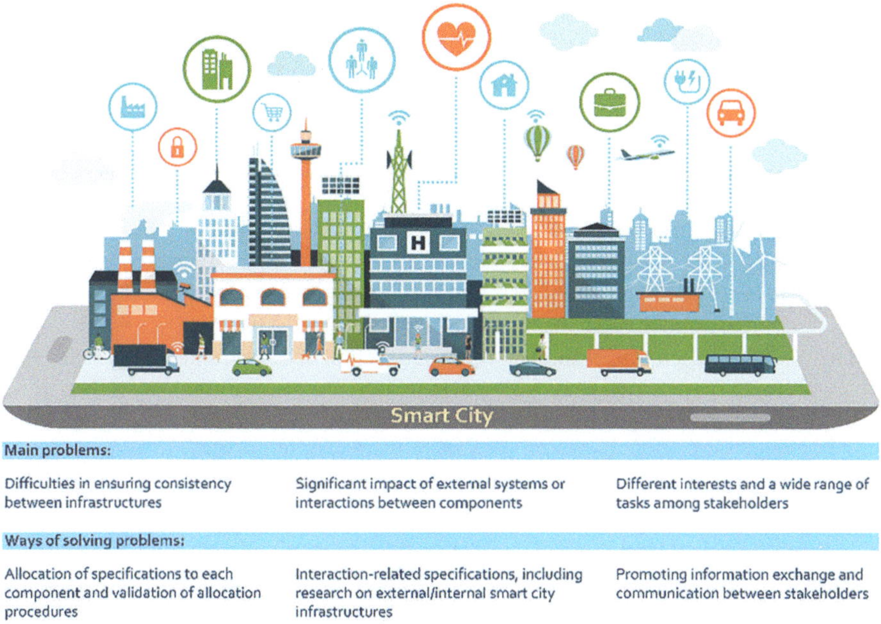

Fig. 2 Possible problems and solutions in smart city

city. Restriction on the use of GIS data exists in 100% of surveyed cities; data are available only to certain organisations.

2 Infrastructure Projects Risks in Smart City

In the phased implementation of smart city infrastructure projects, there is a need to identify risks in order to make decisions about the management of such projects.

Risk clustering of smart city infrastructure projects is carried out according to all stages of the project life cycle and takes into account both systematic risks, e.g. political, financial risks and non-systematic risks, e.g. technical and integration risks, including project interdependencies [11].

According to the research conducted by the McKinsey Global Institute, the world invests about $2.5 trillion annually in infrastructure (transport, utilities, communications). Although significant, this amount is insufficient for the expected sustainable development of states' infrastructures [12].

However, with proper management of infrastructure projects and logistics projects and by maximizing the use of existing infrastructure, it is possible to save up to $1 trillion a year [13].

A significant trend in smart city risk management is the proactive approach or proactive risk management. The approach involves the use of risk analysis and assessment models and tools that project managers can use to identify risks, make decisions, and monitor responses to infrastructure project risks [14, 15].

Because of the multitude of individual projects in an infrastructure project with interlinked elements, including spatial data, high risks of non-implementation [16], decision-making on the implementation of individual projects and their elements by different decision makers, a decision support system for managing the implementation of infrastructure projects is used.

When deploying or maintaining smart city infrastructure projects, the task of visualising the representation of their topological characteristics and certain attributes is realised. The results of the projects are not static; they are in a dynamic state.

The topological characteristics are the geographical coordinates of the elements of smart city infrastructures, as well as their corresponding functional characteristics. Representation of the current state of infrastructure project results is possible in the form of a topological structure with the use of geoinformation systems software [17].

Standard industry and commercially available software tools widely used for project management and control generally use the critical path and Monte Carlo methods, but these methods do not allow the degree of influence of infrastructure projects on each other to be determined, nor are they able to reflect the full structure of project linkages while remaining visually comprehensible to the user.

Evaluation of the results of monitoring the state of implementation of infrastructure projects can be provided on the basis of graph models, e.g. in the form of a weighted undirected graph. The vertices and edges of such a graph will correspond to the spatial data of the actual infrastructure elements [11].

Examples of the most common and widely available graph modelling tools for mass users are yEd, Gephi, Tulip, Cytoscape and Graphviz. A comparison of the main characteristics of the tools is given in Table 1.

To assess the interconnection risk of smart city infrastructure projects, it is advisable to use the following graph analysis methods (Fig. 3) [18]:

- Estimating the *weighted degree centrality* of a vertex in a graph: The weighted degree of centrality is a representation of the total sum of the weights of the edges that are connected to a vertex. The type of centrality quantifies the ability of a vertex to influence other vertices in the graph. Depending on which attributes of an infrastructure project are the weights of the edges, it is possible to determine the significance of an infrastructure object, the risk of its loss, and the level of development of an infrastructure site by using this indicator.
- Calculating the *betweenness centrality* of a vertex in a graph: Represents the calculation of the number of shortest paths between graph vertices that pass through the vertex in question. This type of centrality characterises how important the infrastructure object is in connecting other pairs of objects. It can be used, for example, in communication networks to determine whether existing infrastructure objects can be used.

Table 1 Comparison of graph modelling tools

Parameter	yEd	Gephi	Tulip	Cytoscape	Graphviz
Developer	yWorks GmbH	UTC, The Gephi Consortium	David Auber, Patrick Mary	Institute of Systems Biology	AT&T
Operation system	Linux, Windows, macOS, other that support the Java VM	Linux, Windows, macOS	Linux, Windows, macOS	Linux, Windows, macOS	Linux, Windows, macOS, other *nix
Programming language	Java	Java, OpenGL	C++	Java	C
Distribution	Free	GNU General Public License, Common Development and Distribution License	GNU Lesser General Public License	GNU Lesser General Public License	Eclipse Public License

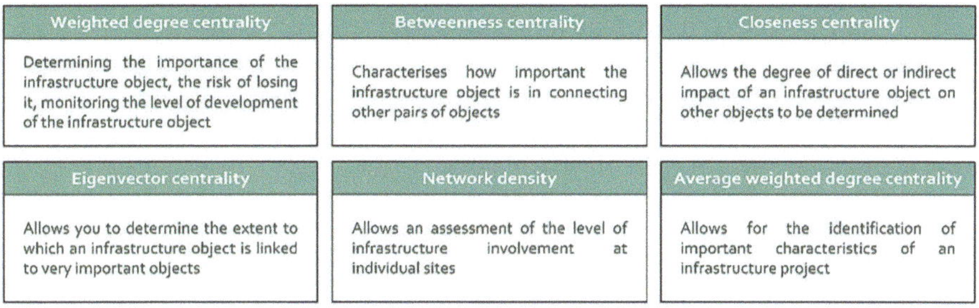

Fig. 3 Numerical methods for analysing infrastructure projects

- Finding the *closeness centrality* of a vertex in a graph: It consists in determining the number of edges on the shortest paths from a vertex of the graph to all other vertices. This type of centrality essentially shows how close a vertex is relatively to the other vertices. In a study of infrastructure projects, this indicator helps to determine the extent to which the infrastructure has a direct or indirect impact on other objects.
- Eigenvector centrality estimation: Allows estimating the degree to which a vertex is connected to other influential vertices (with a high degree of centrality). This type of centrality is used to investigate the extent to which a vertex is connected to very important vertices in the graph. In infrastructure project management, this indicator is important to use when implementing critical infrastructure, security infrastructure to ensure timely preventive action.

- *Network density* estimation: It shows the ratio of the actual number of connected edges in the network to the maximum possible number of edges. This measure of how well connected the graph is has values from zero to one, where zero indicates that the vertices of the graph are completely unconnected and one indicates complete connectivity. In infrastructure projects, this indicator can be used to assess the level of infrastructure involvement at individual sites.
- Determination of the *average weighted degree centrality*: It consists in normalizing the sum of the values of the weighted degree centrality for all vertices of the graph by their number. In project management, this indicator can be used to determine the characteristics of an infrastructure project. For example, the average cost of implementing infrastructure sections, the average capacity of a network section, the average distance between infrastructure facilities, and others depending on how the weights of the edges of the graph are expressed.

Numerical analysis of graph models should be used in procedures to identify the most risky elements of a graph (vertices, branches, sections of the graph) to plan infrastructure deployment and support decision-making by decision-makers.

The use of the proposed risk modelling methods for infrastructure projects, including the use of spatial data and numerical methods for their assessment, will enhance the capacity of decision support systems for those responsible for the implementation of infrastructures in smart cities. An in-depth analysis of numerical methods for risk assessment of interconnected infrastructure projects, proposals for the integration of modelling tools and decision support systems of organizations should be considered as an area for further research on spatial risk data.

Acknowledgement The author would like to thank the editors and reviewers for their careful attention to the chapter and the comments made, which improved its quality. The chapter has been prepared in the framework of applied scientific research conducted by the Bonch-Bruevich Saint-Petersburg State University of Telecommunications in 2022.

References

1. Edinyj reestr informacionnyh resursov i sistem goroda Moskvy (po sostoyaniyu na 20.09.2022). [Cited 20.09.2022]. https://www.mos.ru/dit/documents/proekty-i-otchety/view/350220/ [Links]. (in Russian)
2. Oberascher, M., Rauch, W., Sitzenfrei, R.: Towards a smart water city: a comprehensive review of applications, data requirements, and communication technologies for integrated management. Sustain. Cities Soc. **76** (2022)
3. Visan, M., Negrea, S.L., Mone, F.: "Towards intelligent public transport systems in smart cities", collaborative decisions to be made. Procedia Comput. Sci. **199**, 1221–1228 (2022)
4. Singh, S.K., Pan, Y., Park, J.H.: Blockchain-enabled secure framework for energy-efficient smart parking in Sustainable City environment. Sustain. Cities Soc. **76** (2022)
5. Akindipe, D., Olawale, O.W., Bujko, R.: Techno-economic and social aspects of smart street lighting for small cities – a case study. Sustain. Cities Soc. **84** (2022)
6. Mortaheb, R., Jankowski, P.: Smart city re-imagined: City planning and GeoAI in the age of big data. J. Urban Manag. (2022)

7. Abhishek, K., Kaur, S.: Internet of things (IoT), applications and challenges: a comprehensive review. Wirel. Pers. Commun. **114**(2), 1687–1762 (2020)
8. ISO 37156:2020 – Smart community infrastructures – Guidelines on data exchange and sharing for smart community infrastructures
9. ISO/TR 37152:2016 – Smart community infrastructures – Common framework for development and operation
10. Turek, T., Stępniak, C.: Areas of integration of GIS technology and smart city tools, research findings. Procedia Comput. Sci. **192**, 4681–4690 (2021)
11. Shestakov, A.V., Pletnev, Y.A., Frolova, K.A.: Geoinfomationnyye sistemy v upravlenii i monitoring tekhnogennykh ob 'yektov. Skhemy i QR-ssylki in, SanktPeterburg: SPb.: Tipografiya Lyubavich, 100 p. (in Russian) (2021)
12. McKinsey Global Institute: Bridging infrastructure gaps: has the world made progress? Executive Briefing [Cited 20.09.2022]. https://www.mckinsey.com/capabilities/operations/our-insights/bridging-infrastructure-gaps-has-the-world-made-progress [Links] (2017)
13. McKinsey Global Institute: Infrastructure productivity: how to save $1 trillion a year. Report [Cited 20.09.2022]. https://www.mckinsey.com/capabilities/operations/our-insights/infrastructure-productivity [Links] (2013)
14. Rihar, L., Zuzek, T., Berlec, T., Kusar, J.: Standard risk management model for infrastructure projects. In: Risk Management in Construction Projects. IntechOpen (2019)
15. Jovanovic, J., Mosurovic, M., Beric, I.: Risk as a factor of decision making in projects financing of infrastructure projects. Eur. Project Manag. J. **10**(2), 11–17 (2020)
16. Tripathy, B.K.: Risk assessment in IT infrastructure. In: Security and Privacy from a Legal, Ethical, and Technical Perspective, pp. 1–14. IntechOpen (2020)
17. Shestakov, A.V.: Vvedenie v metodologiu obrabotki geoprostranstvennuh dannuh genotipa telekommunikasei. SPb.: GUAP – Sankt-Peterburg, 325 p. (in Russian) (2016)
18. Gondia, A., Ezzeldin, M., El-Dakhakhni, W.: Dynamic networks for resilience-driven management of infrastructure projects. Autom. Constr. **136** (2022)

A Comparative Analysis of Blockchain-Based Authentication Models for IoT Networks

Tran Duc Le, Truong Duy Dinh, Quoc Khanh Dang, Thi Le Quyen Nguyen, and Ruslan Kirichek

1 Introduction

Recently, one of the most researched topics by academics is the Internet of Things (IoT), which is a term that refers to linking any gadget to the Internet [1]. IoT is a network of physical items outfitted with sensors, software, and other technologies that enable them to communicate and share data with other systems and devices over the Internet. According to the Ericsson Mobility Report, there will be 550 million 5G subscriptions in 2022, and Asia-Pacific will be the second-fastest-growing area for 5G subscriptions in 2022, accounting for 10% of all subscriptions. By 2022, there are expected to be over 29 billion connected devices, with around 18 billion of those being associated with IoT. According to IDC company, the globe will have 41.6 billion linked IoT devices, or "things," by 2025. Besides that growth potential, IoT faces many challenges, such as interoperability, compatibility, limited bandwidth, data complexity, data volume, and especially security issues. The leading IoT security concerns are authentication, authorization, integrity, availability, and privacy [2]. In this research, we focus on the authentication issue in

T. D. Le · Q. K. Dang · T. L. Q. Nguyen
University of Science and Technology – The University of Danang, Danang, Vietnam
e-mail: letranduc@dut.udn.vn; ntlequyen@dut.udn.vn

T. D. Dinh (✉)
Posts and Telecommunications Institute of Technology, Hanoi, Vietnam
e-mail: duydt@ptit.edu.vn

R. Kirichek
The Bonch-Bruevich Saint Petersburg State University of Telecommunications, Saint Petersburg, Russia
e-mail: kirichek@sut.ru

A. A. Abd El-Latif et al. (eds.), *Secure Edge and Fog Computing Enabled AI for IoT and Smart Cities*, EAI/Springer Innovations in Communication and Computing,
https://doi.org/10.1007/978-3-031-51097-7_18

IoT networks because the authentication of users and devices in this network is still dependent on a third party.

Additionally, according to Gartner, the corporate and automotive industries will account for 5.8 billion devices this year, up nearly a quarter in 2019. These demonstrate the significance of IoT devices in the digital world nowadays [1]. Besides, IoT applications such as healthcare, smart home, and agricultural applications demonstrate that IoT has appeared everywhere and in all areas of life. Aside from the rapid development of IoT devices, network security issues are also rising. In reality, IoT implements network technologies and integrates existing network infrastructures (e.g., wireless sensor networks, RFID-based sensor networks, Cloud Computing, and the Internet) [3]. Therefore, security-related issues, as well as challenges in IoT networks, will revolve around these technologies. Specifically, the IoT network's security core involves confirming the legitimacy of a device's identification when it connects to the network. Authentication is the process by which a network decides whether a device/user is authorized to access specified resources. Authentication may be classified into three types: what you know (knowledge), what you have (possession), and who you are (ownership) [2]. The identities of devices must be validated, and conventional IoT authentication depends on Public Key Infrastructure (PKI), which requires a central authority. With an increasing number of devices and users, centralized systems cannot keep up with the system's demands, particularly the authentication process.

As a new and powerful distributed ledger system, blockchain enables a trustless environment that can eliminate reliance on a central authority. Each node uses a consensus method to ensure that stored data is consistent in a blockchain network. As long as honest nodes are in control, the system is safe. They control more hashrate than any other group of cooperative attacker nodes [4]. Regardless of how the attacker enters the system, the data is preserved. Blockchain technology provides suitable options for establishing identification and protecting the security of IoT devices [5–7]. However, the application of blockchain technology to the IoT network is relatively new, and there are still some issues needed to be resolved.

This paper presents an overview and analysis of blockchain-based authentication models for IoT Networks. The remainder of this study is organized as follows: In the next section, the existing authentication methods for the IoT networks are analyzed, and the drawbacks of these methods are considered. Next, we present the notion of blockchain technology and discuss its advantages suitable for authentication in IoT networks. Next, blockchain-based authentication models for IoT networks are analyzed and compared. The last section will provide the conclusions of this paper.

2 Authentication Methods in IoT Networks

The IoT application areas are vast and include e-health, smart cities, smart home, wearables, etc. [8]. As a result, tens of billions of devices will be linked, if not hundreds of billions. These intelligent gadgets will be capable of collecting, analyzing, and even making choices independently of human contact. In such cases, security is of paramount importance, and authentication, in particular, is critical given the potential harm caused by an unauthenticated hostile device in an IoT system [9, 10].

These challenges are "exacerbated" by the resource-constrained nature of IoT devices, which makes conventional communication protocols and security solutions inefficient for IoT. In addition, authentication is a critical prerequisite for IoT [11–13]. Trusted devices connected to an IoT network are critical to the network's proper operation. A single hacked node may be turned hostile, bringing the whole system down or causing disasters. When attackers use the intelligence embedded in homes, automobiles, and energy grids, it may be used to create disastrous situations. Over the last few years, numerous hacking scenarios have revealed the wide range of the damage caused by a security breach, most notably with the development and widespread adoption of IoT applications that manage sensitive data (governmental, industrial, personal, etc.).

Numerous criteria are used to build different IoT authentication models [8–12]. Suppose the authentication model is based on authentication factors such as identity (a piece of information offered by one party to another to establish its legitimacy) or Digital object architecture (DOA). In that case, we can use cryptographic algorithms (hash, asymmetric, symmetric). Physical (fingerprints, hand geometry, retinal scans) and behavioral (keystroke dynamics, voice ID) traits are also used for this type of authentication—biometric-based models.

Besides models based on authentication factors, there are models using tokens to authenticate the network access, for example, token-based (using an identification token provided by a server such as OAuth2 protocol) authentication model, non-token-based (requiring usage of username/password) authentication model. The authentication models are also built based on the procedure of this process itself. In the case of two parties seeking to communicate, if only one will authenticate itself to the other, leaving the other unauthenticated, we have a one-way authentication model. If both parties authenticate one another, there is a two-way authentication model. If a central authority is used to authenticate or help all parties authenticate themselves, we have a three-way authentication model. There are different architectures used in IoT networks. Therefore, each architecture needs its authentication model [14]. For centralized cases, authentication credentials are provided and managed by a server or trusted third party. For distributed cases, a straightforward authentication technique is used between parties. If IoT networks use hierarchical architecture, the authentication process must be carried out by a multi-level approach.

The authentication models in IoT are diverse on different layers, such as the application, network, and perception layers. On the physical layer, the authentication should consider the physical characteristics of the hardware. In that case, there are implicit hardware-based and explicit hardware-based authentication models [8].

Due to the nature of the IoT ecosystem, identification and authentication may be problematic [1]. It is critical to examine identity authentication management in the IoT since various individuals and devices must authenticate through trustworthy services. Additionally, efficient key distribution and maintenance are difficult with IoT devices since they may result in node overhead. Moreover, in the absence of a verified Certificate Authority (CA), additional techniques for verifying cryptographic keys and assuring key transfer integrity are necessary. The blockchain approach may be a good solution in this case.

Moreover, due to the enormous growth in IoT devices and the high level of societal implications of the IoT, such as the vast variety of anticipated uses, anti-counterfeiting has become a significant problem. DOA is a worldwide system that facilitates the production of digital things with unique IDs. Using DOA, any information may be identified and saved using a globally unique identity. The system assigns a unique identification to each IoT device and allows the end user to access device information and specifications through the identifier [15–18].

3 Blockchain Concept

In 2008, blockchain was proposed by Satoshi Nakamoto [4]. A blockchain is formed of blocks that are connected in a chain. Each block includes the block's ID, the preceding block's hash, a nonce, and transaction data. The chain is established by recording the preceding block's hash in each block. The ledger is the name given to this chain [19]. Figure 1 illustrates a chain of blocks in the blockchain.

Each node in the network has its ledger. In the blockchain, consensus techniques validate transactions and update the whole ledger. When a new transaction is recorded to the ledger, all nodes in the network verify the data's correctness and, upon approval, add the new transaction to their ledgers. Each user registers a pair of public and private keys when joining the network. The recording of a transaction performs it. Each user's wallet contains the keys to their account. The blocks were created by miners. Miners are network nodes assigned to create and approve blocks for the blockchain. For an instant, to create a block in Bitcoin [4], the corresponding

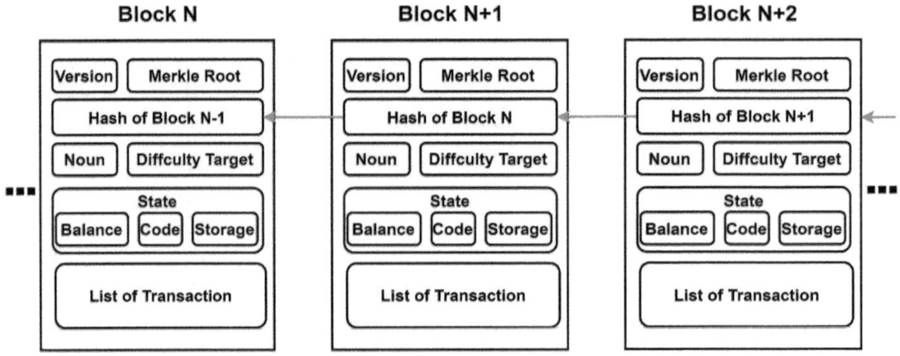

Fig. 1 Chain of blocks in the blockchain

node must solve a difficult puzzle, and whoever solves it first registers their block in the blockchain. Changing an authorized ledger block is expensive and time-consuming.

Blockchain consists of three types: public, private, and consortium. Anyone may participate in a public blockchain's block generation and consensus processes. On a private blockchain, only preapproved nodes can do it. However, a consortium is a hybrid of private and public. Bitcoin and Ethereum are public-type blockchains; Ripple (XRP) (https://ripple.com) and Hyperledger (https://www.hyperledger.org) are examples of private-type blockchains; Quorum (https://quorum.com) and Corda (https://www.corda.net) are consortium-type blockchains.

Starting with Ethereum, a new phrase has emerged in the blockchain world: smart contracts. Smart contracts are executable programs and memories recorded on the blockchain as transactions. Miners execute smart contracts at a fixed fee. It is feasible to invoke the smart contract for network members by knowing the transaction address. Because blockchain is immutable, smart contracts may instill a sense of security. Additionally, smart contracts serve to boost the scalability of blockchain applications.

The three primary pillars [20] upon which blockchain technology is built that have aided in its global acceptance are as follows:

- Decentralization: no one entity retains ownership or control of the data recorded on the blockchain. Each network entity owns this information, and everyone must agree upon any modifications to this chain of knowledge.
- Transparency: anyone checking a basic public blockchain or joining a private blockchain network can observe every transaction and related information.
- Immutability: whenever the block containing a set of transactions is added to the blockchain, altering these transactions is almost infeasible. Blockchain technology accomplishes this characteristic via the use of cryptographic hash functions. Besides, several more aspects that are critical to the operation of any blockchain network include the following:
- Distributed ledger: a blockchain is a decentralized ledger. It means the data recorded in the blockchain is distributed throughout the network's nodes.
- Security: as blockchain technology eliminates the need for a central authority, no organization may alter the network's properties directly to their advantage. Additionally, the system's security is enhanced through encryption and other cryptographic methods.
- Scalability: scalability refers to the blockchain's ability to operate without experiencing low-speed processing times, system bloat, or delays, even on a massive network. Since a blockchain grows in size as it gains popularity and demand, it should be capable of processing millions of transactions per second.

Due to these advantages, using blockchain for the authentication of nodes in an IoT network increases the network's security and transparency.

3.1 Role of Smart Contract

A smart contract acts as an authentication center storing information of users, devices, and relationships between users and devices. In this paper, Smart Contract is built using Solidity. The smart contract is compiled and migrated through Truffle. The compilation process will convert the Solidity code into bytecode and the Application Binary Interface (ABI). The bytecode will be used to deploy to the Ethereum network, while the ABI determines which function in the Smart Contract is called and returns the data in the expected format.

Migration is the process of converting compiled data onto the blockchain network. For Truffle, this process brings the compiled bytecode to a private blockchain network—specifically Ganache. Corresponding to each Smart Contract, an address will be used to identify it on the blockchain. This address is a long string of numbers and characters starting with 0x. Each address is associated with four different fields:

- Nonce: is an integer that is incremented every time the address sends any transaction.
- Balance: is the balance of the Smart Contract. If the user makes a transaction to this address, the balance will increase, and when the transaction is sent to another address, this balance will decrease.
- Code: source code of Smart Contract.
- Data: is where all the storage variables of a smart contract are stored. Memory variables cannot be stored in the blockchain.

3.2 Analysis of Blockchain-Based Authentication Models for IoT Networks

As mentioned above, blockchain has many essential characteristics. They are distributed architecture, transparency, immutability, indestructibility, and fault tolerance. Thanks to these characteristics, blockchain-based authentication represents a novel and appropriate approach to the authentication of IoT networks. Previous IoT network authentication methods were performed by a third-party middleman, who would retain all information and have complete control over the user's authentication.

The fault tolerance of the blockchain enables one or a few nodes to suspend operations without affecting validation. In the blockchain, user credentials are public and stored in blocks. Each user can verify the other by checking whether the user's information already exists in the blockchain and whether the user's access permission is valid.

Two objects must be authenticated in the current IoT networks: users and devices. User authentication aims to determine whether a valid user wants to access the

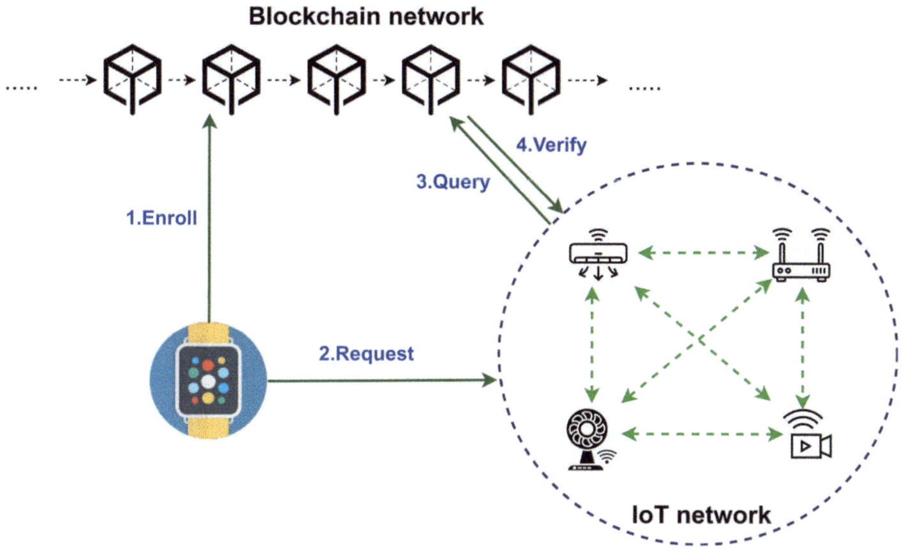

Fig. 2 Authentication model in [21]

resources within the scope of the authorized access to prevent the user from tampering with access to sensitive information.

Device authentication is primarily used to detect which devices are legitimate and malicious. Malicious devices can impersonate ordinary devices by making access requests to the rest of the network to get information that compromises the security of user data. Due to the limitations of the computing power and storage capacity of devices in the IoT network, an appropriate authentication model can assist in protecting devices from unauthorized access by attackers or malicious devices. This section illustrates and analyzes some popular blockchain-based authentication approaches for IoT networks.

In the paper [21], Dongxing Li and Wei Peng proposed a model (Fig. 2) that utilized blockchain technology to create a secure, tamper-proof distributed ledger for IoT devices.

The proposed model assigned a unique ID to each device and stored it on the blockchain, allowing devices to verify one another without the need for a central authority. New devices must register directly on the blockchain network to connect with other devices in the network. Once approved by the blockchain, the device will be able to connect to the IoT network and conduct peer-to-peer (P2P) data exchange with other devices. It enables rapid device-to-device communication as well as rapid authentication. However, for devices that are unable to self-protection, this might threaten the IoT network and P2P communication devices. This model mainly authenticates devices in the network, so it does not take into account the case of user authentication using devices.

Additionally, a data security system was developed by hashing important data into the blockchain, where any state changes to the data are quickly noticed. The authors simulated the system's deployment using Hyperledger Fabric to evaluate the model's effectiveness. The Raspberry Pi served as a network node, and the nodes communicated through an ad-hoc network.

Randa Almadhoun and Maha Kadadha [22] presented an authentication model that is utilized in combination with fog nodes, which are computing systems located closer to devices, hence reducing the processing power required by network devices. However, the administrator must verify the network's devices. It makes network expansion difficult due to device authentication becoming more challenging as the number of devices increases. Besides, the smart contract provides the device-to-fog node mappings and a list of user authentications. In this model, a device is only associated with a fog node, but a fog node can be connected with multiple devices. When a user wants to access a specific IoT device, it must submit a request to the smart contract for permission to access this IoT device. After obtaining permission to access the target device, the user must send an access request to the fog node. The fog node authenticates the user and grants or refuses user access. The author deployed this approach on the Ethereum network and smart contract written in the Solidity programming language. Figure 3 shows the user authentication scheme of IoT devices using blockchain-enabled fog nodes.

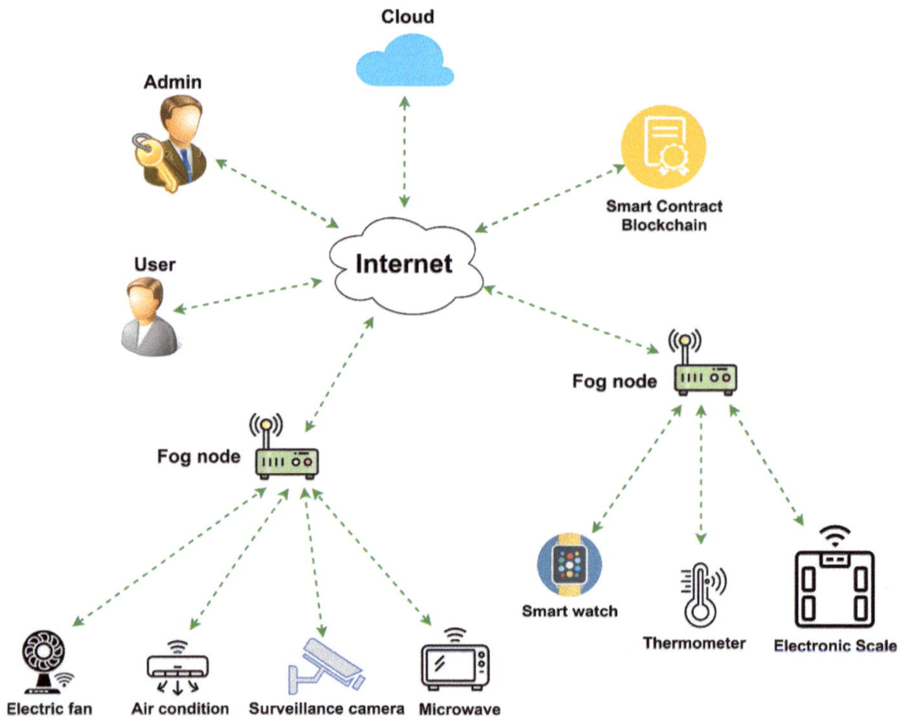

Fig. 3 Authentication model in [22]

Mohamed Tahar Hammi in [23] proposed a novel security authentication blockchain-based mechanism for nodes in a wireless sensor network (WSN), a subset of the Internet of Things (IoT). This method is called BCTrust, based on the principle "The friend of my friend is my friend." It implies that once a device is authenticated in one of the WSN's clusters, it is trusted and accepted by all other clusters. The blockchain guarantees that the recorded data is accessible to and unmodifiable by all participating nodes. Additionally, each device in the network is managed by a CPAN (Personal Area Network Coordinator). When a device connects to the CPAN, a connection record is created on the blockchain. When a device switches to another CPAN, that new CPAN checks whether the device is already connected to the network or is a member of a prior CPAN. If the device is found on the blockchain, the current CPAN will contact the device's prior CPAN and undertake a decentralized migration to the current CPAN. Then, it updates the blockchain with the device's updated information. Although there is no computation to help devices in the network, this model offers a possible solution for the mobility of devices. The model is deployed on the Ethereum Blockchain. The detailed operation of the model is described step by step as in Fig. 4.

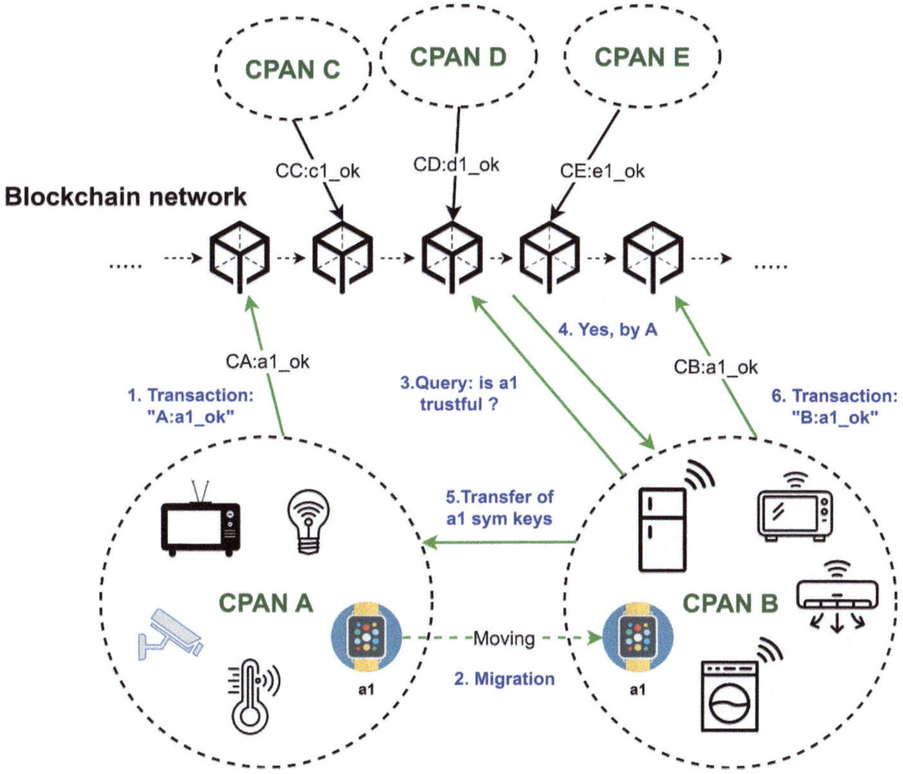

Fig. 4 Authentication model in [23]

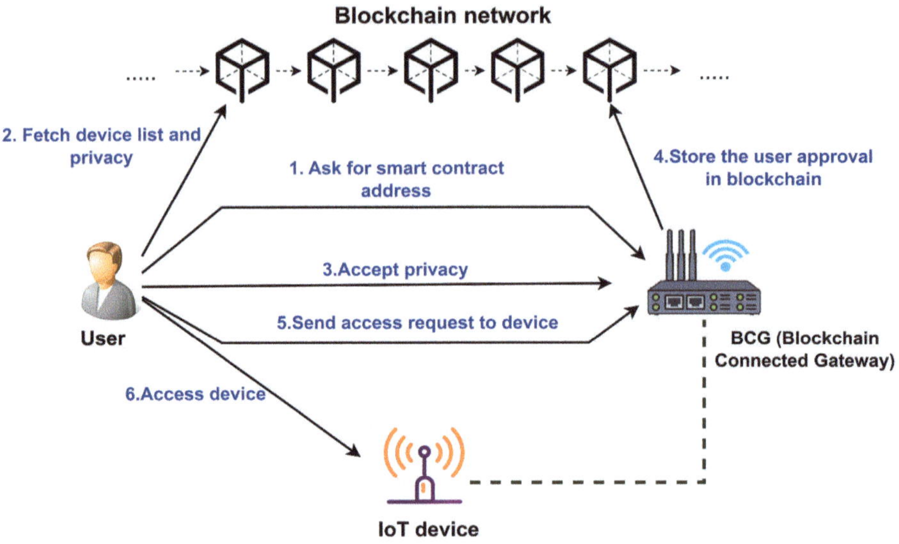

Fig. 5 Authentication model in [19]

Mostafa Yavari in [19] proposed a device and user authentication model based on blockchain technology and gateways. In this model, a user submits a request to the gateway for a blockchain smart contract address to get the user and privacy policy information. If approved, the user will submit another request to the gateway, which will store the user and device data on the blockchain (Fig. 5). Apart from user authentication, devices must be signed by an administrator or another device manager (Fig. 6). This model has three primary elements: the device, the user, and the blockchain-connected gateway. Devices are network nodes that exchange information or resources under a device policy (smart contract) with other users, devices, or people who want to utilize the devices' information or resources.

According to the policy, the Blockchain-connected gateway will assess the user's eligibility (permissions and authentication) to use the device's information and resources. Once approved, the user has complete control of the gadget. The authors deploy the system on the local Ethereum blockchain network using Nodejs.

Moreover, in [24], H. Albreiki et al. proposed a model that authenticates both devices and users in the IoT network. Through the blockchain, users are authenticated and granted access to the device data. The administrators manage access for users and devices. Before gaining data access, users must submit a request to the blockchain network for authentication and authorization. The IoT Data Storage Layer stores data from IoT devices using various storage services/protocols such as Message Queuing Telemetry Transport (MQTT) brokerage, Constrained Application Protocol (CoAP), etc. The paper used a new technology called Oracle, an intermediary to bring external data to the blockchain, providing data for the entire

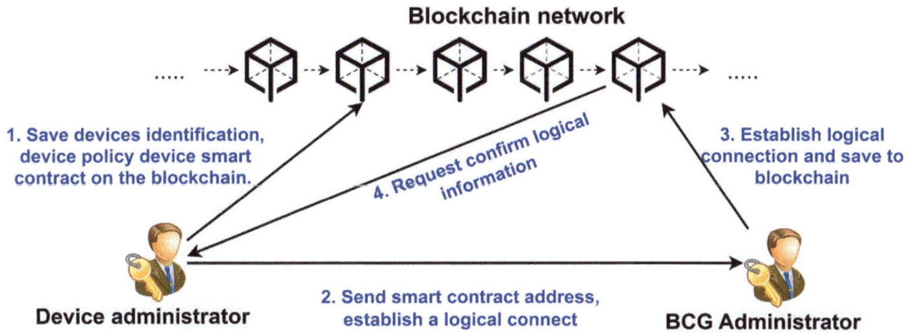

Fig. 6 Sequence diagram for device authentication when storing data

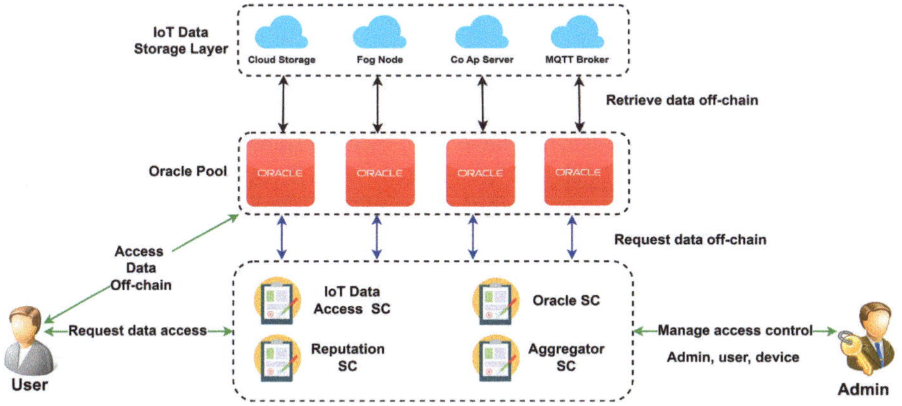

Fig. 7 Authentication model in [24]

system. The author evaluates and deploys the model on the Ethereum blockchain and analyzes the states and costs at each step in the model (Fig. 7).

In brief, all of these models must store user or device data in the blockchain to act as a method of user and device authentication across the network. Each user or device in the network communicates using an asymmetric key pair to authenticate their identity. Because user data is recorded on the blockchain, objects cannot be accessed without a clear identity. Only the private key owner with the appropriate user credentials may access it as a user. Thanks to the blockchain, user authentication data will be impossible to hack and forge. However, anyone may access and authenticate if the device or user has a public key, which is a huge advantage of blockchain technology.

Additionally, scalability and fault tolerance are increased when authentication is based on blockchain technology. Authentication may be accomplished as long as users access the blockchain network. Because the blockchain is spread, the danger of a single point of failure is eliminated, making it more difficult for attackers to attack. The final comparison of the considered models is shown in Table 1.

Table 1 The comparison of the blockchain-based authentication models in IoT networks

Model	Device authentication	User authentication	Cloud computing	Use key pair for authentication	Storage of device data on the blockchain	Blockchain platform	Human management	P2P
[21]	x			x	x	Hyperledger Fabric		x
[22]	x		Fog node			Ethereum	x	
[23]	x			x	x	Ethereum		
[19]	x	x		x	x	Ethereum		
[24]	x	x	Oracle	x		Ethereum	x	

4 Conclusions

This paper reviews the ten popular persistence techniques implemented by almost all modern malware. These techniques allow the malware to survive system reboots and shutdowns. The techniques that have been mentioned are:

In summary, we discussed the fast growth of IoT in recent years and the challenges associated with authenticating users and new devices on the IoT network. Due to its transparency, indestructibility, and fault tolerance, blockchain technology has recently been used in many areas. Thus, the use of blockchain in an IoT network represents a new approach to user and device authentication since it resolves the issues associated with traditional authentication. This paper analyzed previous studies on this topic. From the above analysis, we have some conclusions as follows:

- Using blockchain for authentication in IoT networks is feasible and effective.
- Not all the existing models cover user and device authentication in IoT networks simultaneously.
- The combination of user and device authentication is also an open challenge for future research.
- The issue of transaction distribution or execution time needs to be addressed in future research.
- Using cloud computing is a good way to support the operation of authentication. However, it should be considered the cost of this solution.

Acknowledgments The study was financially supported by the Russian Science Foundation within of scientific project No. 22-49-02023 "Development and study of methods for obtaining the reliability of tethered high-altitude unmanned telecommunication platforms of a new generation."

References

1. Janarthanan, T., Bagheri, M., Zargari, S.: IoT forensics: An overview of the current issues and challenges. In: Digital Forensic Investigation of Internet of Things (IoT) Devices, pp. 223–254 (2021)
2. Weaver, A.C.: Biometric authentication. Computer **39**(2), 96–97 (2006)
3. Andrea, I., Chrysostomou, C., Hadjichristofi, G.: Internet of things: Security vulnerabilities and challenges. In: 2015 IEEE Symposium on Computers and Communication (ISCC), pp. 180–187 (2015)
4. Nakamoto, S.: Bitcoin: A Peer-To-Peer Electronic Cash System. Decentralized Business Review, p. 21260 (2008)
5. Yu, Y., Li, Y., Tian, J., Liu, J.: Blockchain-based solutions to security and privacy issues in the internet of things. IEEE Wirel. Commun. **25**(6), 12–18 (2018)
6. Panarello, A., Tapas, N., Merlino, G., Longo, F., Puliafito, A.: Blockchain and IoT integration: a systematic survey. Sensors **18**(8), 2575, 1–37 (2018)
7. Dorri, A., Kanhere, S.S., Jurdak, R., Gauravaram, P.: LSB: a lightweight scalable blockchain for IoT security and anonymity. J. Parall. Distrib. Comput. **134**, 180–197 (2019)
8. El-Hajj, M., Fadlallah, A., Chamoun, M., Serhrouchni, A.: A survey of internet of things (IoT) authentication schemes. Sensors **19**(5), 1141 (2019)
9. Barreto, L., Celesti, A., Villari, M., Fazio, M., Puliafito, A.: An authentication model for IoT clouds. In: 2015 IEEE/ACM International Conference on Advances in Social Networks Analysis and Mining (ASONAM), pp. 1032–1035 (2015)
10. Tapas, N., Merlino, G., Longo, F.: Blockchain-based IoT-cloud authorization and delegation. In: 2018 IEEE International Conference on Smart Computing (SMARTCOMP), pp. 411–416 (2018)
11. Vishwakarma, L., Das, D.: SmartCoin: a novel incentive mechanism for vehicles in intelligent transportation system based on consortium blockchain. Vehic. Commun. **33**, 100429, 1–11 (2022)
12. Wu, L., Du, X., Wang, W., Lin, B.: An out-of-band authentication scheme for internet of things using blockchain technology. In: 2018 International Conference on Computing, Networking and Communications (ICNC), pp. 769–773 (2018)
13. da Camara, J.: Bubblechain: An IoT Authentication System. Master Thesis, Delft University of Technology, 2021
14. Saadeh, M., Sleit, A., Qatawneh, M., Almobaideen, W.: Authentication techniques for the internet of things: A survey. In: 2016 Cybersecurity and Cyberforensics Conference (CCC), pp. 28–34 (2016)
15. Sazonov, D., Kirichek, R., Borodin, A.: Implementation of authentication and authorization system based on digital object architecture. In: 2019 11th International Congress on Ultra Modern Telecommunications and Control Systems and Workshops (ICUMT), pp. 1–4. IEEE, Piscataway (2019)
16. Albahri, M., Kirichek, R., Ateya, A.A., Muthanna, A., Borodin, A.: Combating counterfeit for IoT system based on DOA. In: 2018 10th International Congress on Ultra Modern Telecommunications and Control Systems and Workshops (ICUMT), pp. 1–5. IEEE, Piscataway (2018)
17. Al-Bahri, M., Yankovsky, A., Kirichek, R., Borodin, A.: Smart system based on DOA & IoT for products monitoring & anti-counterfeiting. In: 2019 4th MEC International Conference on Big Data and Smart City (ICBDSC), pp. 1–5. IEEE, Piscataway (2019)
18. Sazonov, D., Kirichek, R.: Digital object architecture as an approach to identifying internet of things devices. In: International Conference on Distributed Computer and Communication Networks, pp. 597–611. Springer, Cham (2019)
19. Yavari, M., Safkhani, M., Kumari, S., Kumar, S., Chen, C.M.: An improved blockchain-based authentication protocol for IoT network management. Secur. Commun. Netw. **2020**, 1–16 (2020)

20. Kaur, S., Chaturvedi, S., Sharma, A., Kar, J.: A research survey on applications of consensus protocols in blockchain. Secur. Commun. Netw. **2021**, 1–22 (2021)
21. Li, D., Peng, W., Deng, W., Gai, F.: A blockchain-based authentication and security mechanism for IoT. In: 2018 27th International Conference on Computer Communication and Networks (ICCCN), pp. 1–6 (2018)
22. Almadhoun, R., Kadadha, M., Alhemeiri, M., Alshehhi, M., Salah, K.: A user authentication scheme of IoT devices using blockchain-enabled fog nodes. I:n 2018 IEEE/ACS 15th International Conference on Computer Systems and Applications (AICCSA), pp. 1–8 (2018)
23. Hammi, M.T., Bellot, P., Serhrouchni, A.: BCTrust: A decentralized authentication blockchain-based mechanism. In: 2018 IEEE Wireless Communications and Networking Conference (WCNC), pp. 1–6 (2018)
24. Al Breiki, H., Al Qassem, L., Salah, K., Rehman, M.H.U., Sevtinovic, D.: Decentralized access control for IoT data using blockchain and trusted oracles. In: 2019 IEEE International Conference on Industrial Internet (ICII), pp. 248–257 (2019)

Development of Determining a Wireless Client Location Method in the IEEE 802.11 Network to Ensure the IT Infrastructure Security

Drepa Vladislav, Kovtsur Maxim, Minyaev Andrey, Konovalova Viktoria, and Kuzmina Olga

1 Introduction

Information and analytical security systems are used to protect critical wireless networks. Such systems can be wireless intrusion detection systems (Cisco Mobility Service Engine (MSE), wireless Intrusion Prevention System (IPS), etc.). However, not all systems allow you to determine the physical location of the intruder or it requires additional licensing for this functionality awork. The problem has received particular relevance in the context of a growing trend toward import software substitution, as well as the need for independence from specific equipment supplier.

Wireless client location services can be used for information security purposes in the following scenarios:

- Searching for an intruder attacking the company's IT infrastructure via a wireless network.
- Localization of unauthorized access points and other unauthorized devices without using expensive specialized equipment.
- Differentiation of access rights to system resources and IT infrastructure based on location (for example, to deny working with personal data outside the workplace).
- A wireless client movement control (for example, an employee).

D. Vladislav (✉)
The Bonch-Bruevich Saint-Petersburg State University of Telecommunications, St. Petersburg (SPbSUT), St. Petersburg, Russian Federation

K. Maxim · M. Andrey · K. Viktoria · K. Olga
SPbSUT, St. Petersburg, Russian Federation

A. A. Abd El-Latif et al. (eds.), *Secure Edge and Fog Computing Enabled AI for IoT and Smart Cities*, EAI/Springer Innovations in Communication and Computing,
https://doi.org/10.1007/978-3-031-51097-7_19

1.1 General Principles of a Wireless Client Positioning in an IEEE 802.11 Network

Wi-Fi technology provides an additional opportunity to determine the client location for marketing purposes (for example, keeping statistics of customers visits to certain trading floors of the shopping center), in order to navigate or IT infrastructure security ensuring [1, 2].

The wireless client's positioning is carried out by methods that can be separated into two main types: triangulation and trilateration methods. All of these methods require the use of at least three scanning devices. Such devices can be as special scanners or adapters in network monitoring mode. In this chapter, such devices are called sensors.

The triangulation method [3–6] uses the angle of arrival (AoA) principle. This is determining the client location by the angle of signal incidence on the receiver surface. This method allows to achieve high accuracy (2–5 m). However, it requires the mandatory use of specialized expensive antennas or devices equipped with these antennas, such as the Cisco Aironet 3600 or Cisco Aironet 3700. These access points are equipped with a special massive of 32 antennas, each of which receives a signal differently than its neighbor one. The built-in software allows, based on the collected data, to calculate the angle at which the signal comes.

The most frequently used methods for determining the wireless client's location are trilateration methods [6], whose general positioning principle is shown in Fig. 1.

These methods are based on calculating three or more distances from scanning devices to a wireless client. These distances will then be used as radii, with the help of which the desired client location is determined. This location is defined as the intersection point of three circles with radii corresponding to the obtained distances. In this case, it is assumed that the distances have been calculated with perfect accuracy. Otherwise, the desired client location would be represented as the area of intersection of three circles.

In Wi-Fi networks, the following methods are used to determine the distance between the receiver (sensor) and transmitter (desired client) [4–8]:

- Received signal strength indicator (RSSI) – determination of the distance to the object due to the received signal strength indicator. The higher the power, the closer the signal source and vice versa.
- Time of flight (ToF) – determination of the distance to the object by the time of signal passage from the transmitter to the receiver.
- Time difference of arrival (TDoA) – determination of the distance to the signal source due to the difference in the time of signal arrival at different receivers (three or more receivers are required).

Determining distances using RSSI does not require specialized equipment or large computing power, does not impose special restrictions on the scanning device placement, and can also take into account obstacles in the signal path. This method allows to determine the distance with an accuracy of 2 m in ideal conditions.

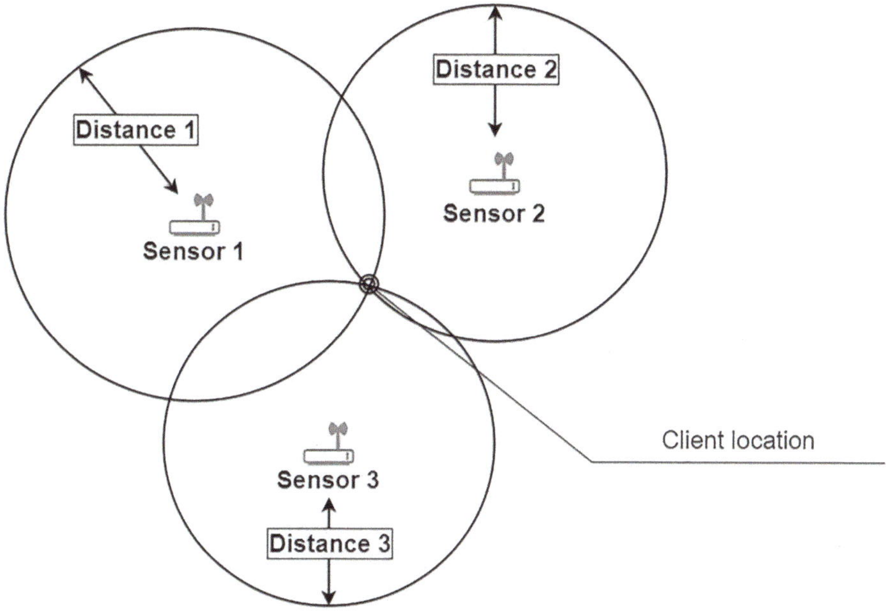

Fig. 1 The general principle of determining the client location by trilateration methods

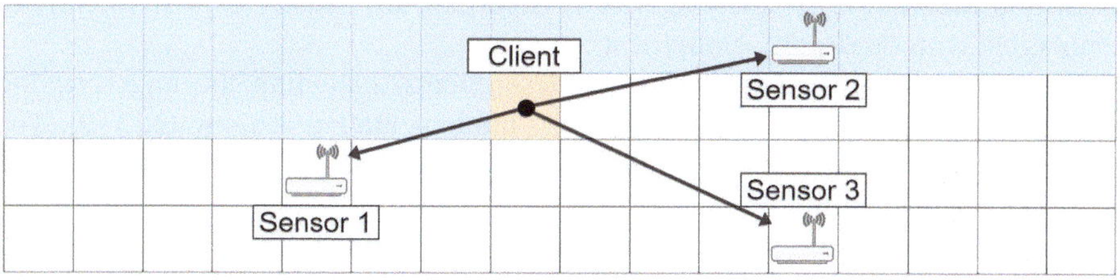

Fig. 2 The fingerprinting method

1.2 Existing Trilateration Methods Based on RSSI

Trilateration methods have many different implementations [4–6, 9, 10]. One of the most common is real-time locating systems (RTLS) [11, 12]. RTLS is real-time positioning based on incoming power information from each scanning device (Fig. 2). This method can be used together with WSM (wireless security module) monitoring modules in Cisco modular APs, as well as in Cisco WLC 2504 and Cisco Aironet 3602 series APs. The main features of this method are high positioning speed, but relatively low accuracy (5–10 m) and the impossibility of accurately accounting for obstacles in the signal path.

The approximation method [9] is similar to RTLS; however, the coordinate of the desired device location is considered to be the coordinate of the sensor that detect this device with the highest power. The approximation method is the easiest to implement and the most inaccurate.

The fingerprinting method [9, 10, 13, 14] (Fig. 2) is based on radio fingerprints. It involves calculating the power for each alleged location of the client, that is, getting an answer to the question "what power would the sensors show if the desired client was at this point." The positioning procedure is reduced to comparing the incoming RSSI data with pre-calculated ones. The fingerprinting method is more resource intensive, but allows for higher positioning accuracy (2–10 m), and can also more accurately take into account additional obstacles in the signal path from the transmitter to the receiver.

In this chapter, the determination of the wireless client location in an IEEE 802.11 family network will be carried out using the fingerprinting method.

2 Problematics

This issue can be considered relevant, due to the presence of a fairly large number of publications related to positioning in Wi-Fi networks over the past 5 years. In many publications, this topic is disclosed in a very abstract way, and the case of the presence of obstacles (walls) in the signal path is not taken into account or partially taken into account. Also, existing vendor solutions are usually expensive and not compatible with third-party equipment.

Various methods for positioning wireless clients, including the fingerprinting method are described, but the details of its implementation are omitted [10]. The proposed implementation of this method does not completely take into account the walls in the signal path [13, 15].

This chapter provides practical recommendations for the third-party equipment-compatible deployment of a system for determining the wireless client's location in networks of the IEEE 802.11 family of standards, taking into account the presence of obstacles in the signal path.

3 Materials and Methods

This work uses a Wi-Fi network, an ALFA Wi-Fi USB adapter, a microcomputer with a WLAN adapter based on a Realtek chipset in three pieces, Python 3.9 and the development environment, and a MySQL database.

4 Results

This chapter discusses the key points in the practical implementation of the method for determining the wireless client location.

4.1 Wireless Client Positioning Method

Figure 3 is a schematic representation of the wireless client positioning method.

To determine the wireless client location:

1. Place at least three sensors on the object.
2. Create an object's map with reference to coordinates, indicating the points where the sensors are located.
3. Generate a power map based on the object's map.
4. Save the calculated power map to the database (DB).
5. Get the value of the signal strength (RSSI) from the client on the facility territory.
6. Compare the power received from the sensors with the calculated power from the map.
7. Display a list of coordinates, if any are found.

Each of the above points will be discussed further.

Figure 4 presents the concept of the wireless client positioning method.

On the left in Fig. 4, the process of power map calculated with the subsequent saving of the results to the database is presented; on the right – the process of processing information received from sensors; and in the center – the process of positioning the wireless IEEE 802.11 client.

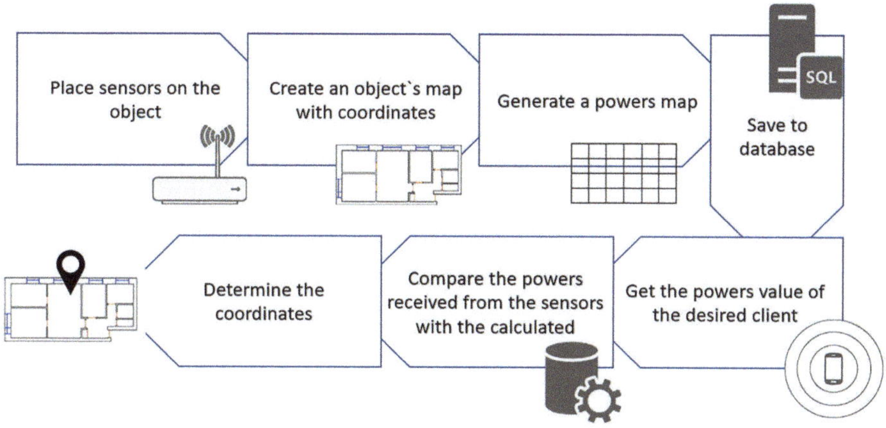

Fig. 3 Wireless client positioning method

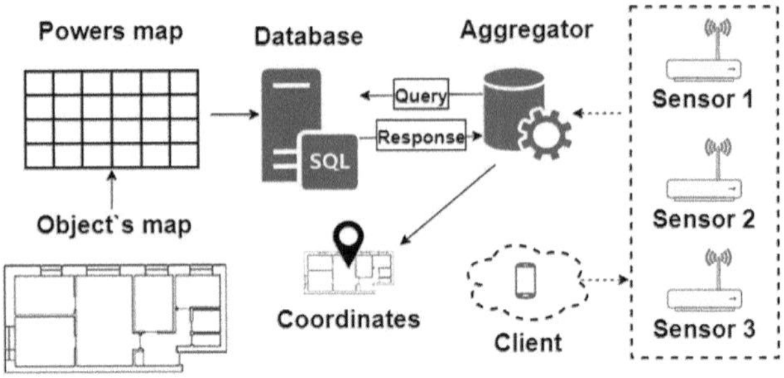

Fig. 4 Concept of the wireless client positioning method

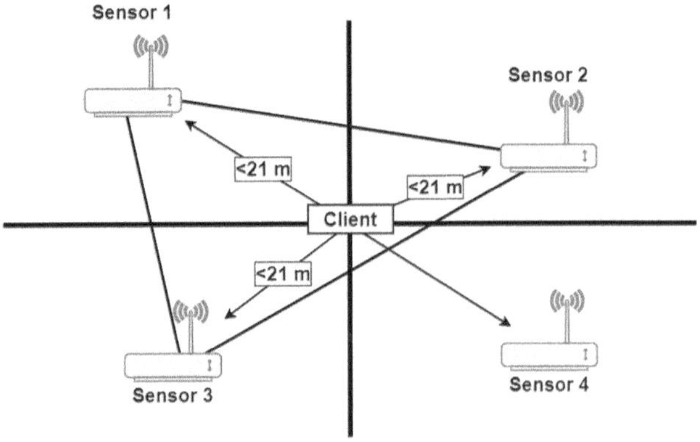

Fig. 5 Sensor placement rule

4.2 *Placement of Sensors on the Object*

For the correct operation of the system, it will be necessary to place the sensors according to the following rule [16], as shown in Fig. 5:

1. If some area is divided into four quadrants, then each such quadrant must contain at least one sensor.
2. The number of sensors must be at least four.
3. At each point in space in the room, the client device must be within a radius of 21 m from at least three sensors.
4. Any three sensors closest to each other taken from this area should form an equilateral triangle (or a triangle close to equilateral), if it is assumed that its vertices correspond to the coordinates of the sensors.

4.3 *Object's Map Creation and Power Map Generation*

Figure 6 shows an object's map (office space). Passive scanning devices – sensors – are placed on this territory. The placement of the sensors is for demonstration purposes only and is not a guide to performing a similar task for this or similar object. The sensors' coordinates and its antenna gain, the coordinates of the walls'

Fig. 6 The office space map with indicated sensor locations

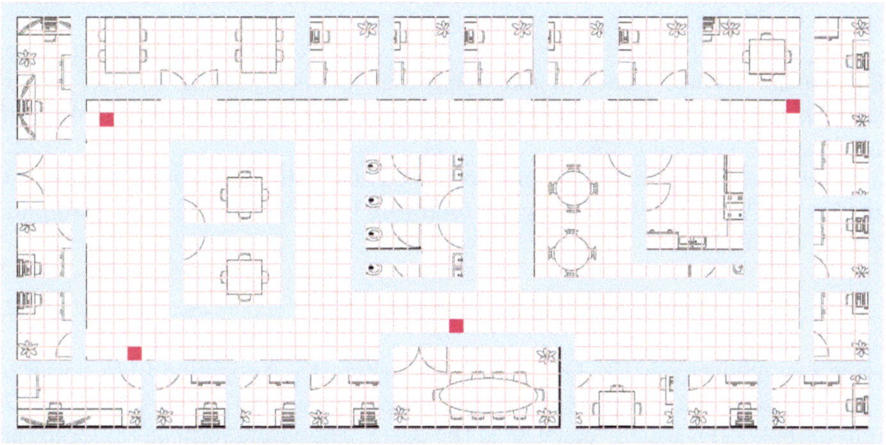

Fig. 7 Office space map divided into cells

location, the thickness of each wall, and the signal attenuation introduced by it for the 2.4 and 5 GHz frequency bands should be entered into the database at the stage of preparing the objec's map.

This map is divided into cells (Fig. 7; here, blue cells are walls and bright pink cells are sensors). Color separation of cells is used for demonstration purposes only and does not affect the properties of these cells in any way. The cell size depends on the area of the object, the number of obstacles for the signal to pass through the object's territory, and the desired calculation accuracy: 0.25 m, 0.5 m, and 1 m. It is advisable to choose larger cells for a large object with a large amount of free space.

For each such cell, the expected RSSI of each of the sensors is calculated for two frequency bands: 2.4 and 5 GHz. The RSSI should not be lower than −85 dBm.

A signal with RSSI below the specified limit should be considered invisible. Such cells are assigned no an unattainable value of −100 dBm. The results obtained are recorded in the database [17].

To fill the database with the values of the received signal strength indications, the following formula is used:

$$\begin{aligned} \mathrm{RSSI} = \mathrm{Pw}_{\mathrm{Tr}} - \mathrm{CblL}_{\mathrm{Tr}} + \mathrm{AntG}_{\mathrm{Tr}} - \mathrm{FSPL} - \\ - \mathrm{ObsAtten} + \mathrm{AntG}_{\mathrm{Rs}} - \mathrm{CblL}_{\mathrm{Rs}}, \end{aligned} \tag{1}$$

where RSSI – received signal strength indication, dBm; $\mathrm{Pw}_{\mathrm{Tr}}$ – transmitter signal power, dBm; $\mathrm{Cbl\_L}_{\mathrm{Tr}}$ and $\mathrm{Cbl\_L}_{\mathrm{Rs}}$ – transmitter/receiver cable loss, dB; $\mathrm{Ant\_G}_{\mathrm{Tr}}$ and $\mathrm{Ant\_G}_{\mathrm{Rs}}$ – transmitting/receiving antenna gain, dBi; FSPL – free space path loss, dB; and Obs_Atten – obstacle attenuation, dB.

In formula (1), the $\mathrm{Pw}_{\mathrm{Tr}}$ value usually does not exceed 20 dBm for the 2.4 GHz band and 23 dBm for the 5 GHz band. The values of $\mathrm{Cbl\_L}_{\mathrm{Tr}}$ and $\mathrm{Ant\_G}_{\mathrm{Tr}}$ are taken as 0 dB and 3 dBi, respectively. The values of $\mathrm{Ant\_G}_{\mathrm{Rs}}$ and $\mathrm{Cbl\_L}_{\mathrm{Rs}}$ must be consulted with the equipment used for specification.

Obs_Atten [18] obstacle attenuation coefficient should be viewed in the tables showing the average values of the signal change when passing through various materials for each frequency range [19]. This coefficient may be given in dB × m and must be converted to dB.

In formula (1), special attention should be paid to the free space path loss calculation. FSPL is calculated using the following formula, reduced to a logarithmic format [14, 20, 21]:

$$\mathrm{FSPL} = 20 \times \lg(\mathrm{f}) + 10 \times \mathrm{D} \times \lg(\mathrm{d}) + \mathrm{p} - 24, \tag{2}$$

where f – frequency, MHz; D – attenuation factor; d – destination from sensor to calculated point, m; and p – penetration factor. The value of the parameter D [22] from formula (2) is usually equal to 2. The coefficient p from formula (2) for the air medium is equal to 1.

4.4 *Saving Data to a Database*

This method does not prescribe the use of a specific database, nor does it regulate the storage structure of the recorded data.

In this chapter, a MySQL database is used. The example of the power map table is shown in Fig. 8.

id	x	y	x_m	y_m	rssi4_24	rssi4_5	dist4	rssi5_24	rssi5_5	dist5	rssi6_24	rssi6_5	dist6
202	41	20	10.25	5	-55.1	-47.6	5.1	-77.4	-71	4.5	-51.9	-41.8	6.9
203	42	20	10.5	5	-55.7	-48	5.2	-79.2	-72.4	4.7	-52.7	-42.4	7.1
204	43	20	10.75	5	-56.3	-48.4	5.4	-81	-73.8	5	-53.4	-42.8	7.3
205	10	21	2.5	5.25	-50.9	-40.6	7	-52.2	-47.2	3.8	-55	-48.6	4.5
206	11	21	2.75	5.25	-50	-40	6.8	-50.7	-46.2	3.6	-54.3	-48.2	4.4
207	12	21	3	5.25	-49.2	-39.4	6.6	-49.2	-45.2	3.4	-53.3	-47.4	4.3
208	13	21	3.25	5.25	-48.3	-38.8	6.5	-47.2	-43.8	3.2	-52.1	-46.4	4.2
209	14	21	3.5	5.25	-47.5	-38.2	6.3	-45.4	-42.6	3	-50.7	-45.2	4.1
210	15	21	3.75	5.25	-46.6	-37.6	6.1	-43.4	-41.2	2.8	-48.6	-43.2	4
211	16	21	4	5.25	-46	-37.2	5.9	-41.5	-40	2.6	-45.5	-40.2	4
212	17	21	4.25	5.25	-48.5	-40	5.8	-39.8	-39	2.4	-57.6	-52.4	3.9
213	18	21	4.5	5.25	-43.1	-34.8	5.6	-31.6	-31.4	2.2	-68.7	-63.6	3.9
214	19	21	4.75	5.25	-43.1	-35	5.4	-30.9	-31.4	2	-39.5	-34.4	3.9
215	20	21	5	5.25	-43	-35.2	5.3	-30.3	-31.4	1.9	-68.7	-63.6	3.9
216	21	21	5.25	5.25	-39	-31.4	5.2	-29.7	-31.4	1.8	-53.4	-48.2	3.9
217	22	21	5.5	5.25	-38.8	-31.4	5	-29.2	-31.4	1.7	-37.7	-32.4	4
218	23	21	5.75	5.25	-38.6	-31.4	4.9	-28.9	-31.4	1.6	-37.8	-32.4	4
219	24	21	6	5.25	-38.4	-31.4	4.8	-28.8	-31.4	1.6	-37.9	-32.4	4.1
220	25	21	6.25	5.25	-38.2	-31.4	4.7	-28.9	-31.4	1.6	-38.1	-32.4	4.2

Fig. 8 Example of power map (here, mapid – id of the powers map; x, y – cell coordinates in conventional units; and x_m, y_m – cell coordinates in meters, RSSI [sensor number]_24 and RSSI [sensor number]_5 – RSSI values for 2.4 and 5 GHz bands for the sensor)

4.5 Client Device Discovery

Figure 9 shows the client device discovery scheme. This device exchanges data packets in a certain Wi-Fi network of the IEEE 802.11 family of standards [23–32].

Sensors intercept these data packets [33, 34], forming network dumps, analyze them, and calculate the average power values of the selected packets. The example of power averaging using the Wireshark network traffic analyzer is shown in Fig. 10.

The calculated average powers are transmitted to the aggregator [35, 36]. The aggregator is a control device that can be represented as a microcomputer (for example, Raspberry Pi) or a server. Next, the process of positioning the client on the map starts.

4.6 Wireless Client Positioning

To position a wireless client, it is enough to search the database to get a list of points of its probable location. It is necessary to select such points at which the previously calculated power for each sensor deviates from the newly obtained powers by no more than 3 dB. For example, if sensors 1, 2, and 3 received a signal in the 2.4 GHz band with average powers of -39, -45, and -54 dBm, respectively, then it is necessary to select points from the database whose powers for this frequency range will be equal to -39 ± 3, -45 ± 3, and -54 ± 3 dBm, respectively. Moreover,

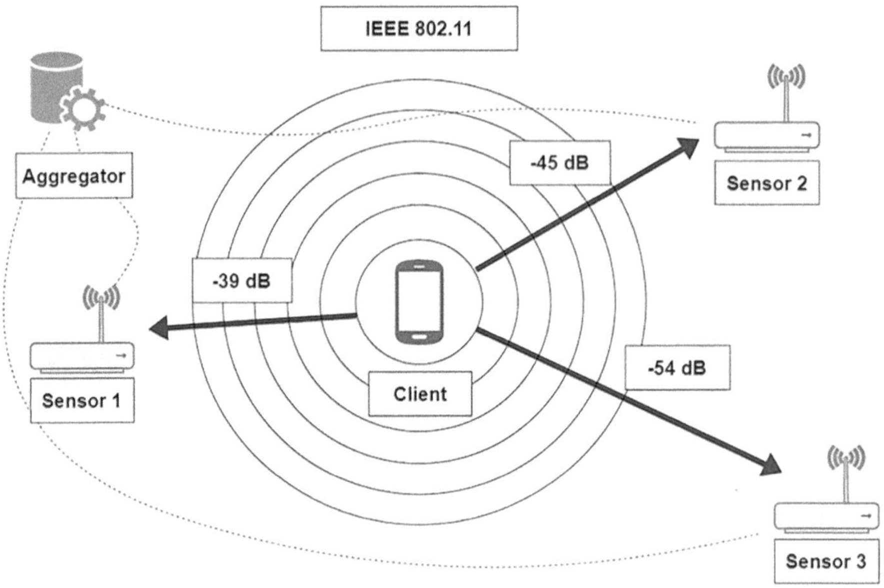

Fig. 9 The client device discovery scheme

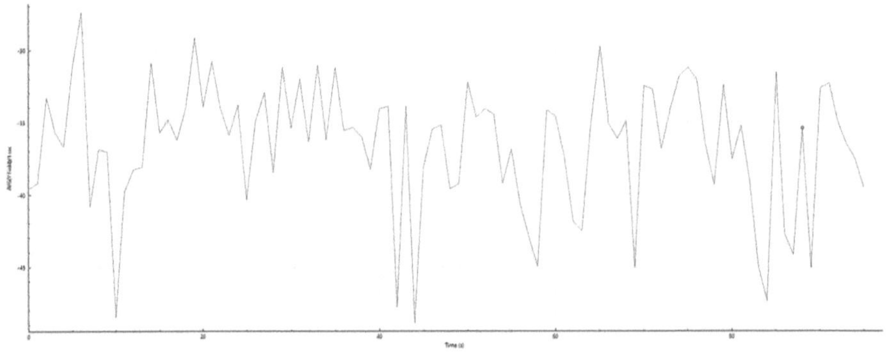

Fig. 10 Example of power averaging at 1 s interval using Wireshark (vertically – power in dBm, horizontally – time in sec)

according to the accuracy of the location of these points in a certain area, one can judge the probability of finding the desired client in it.

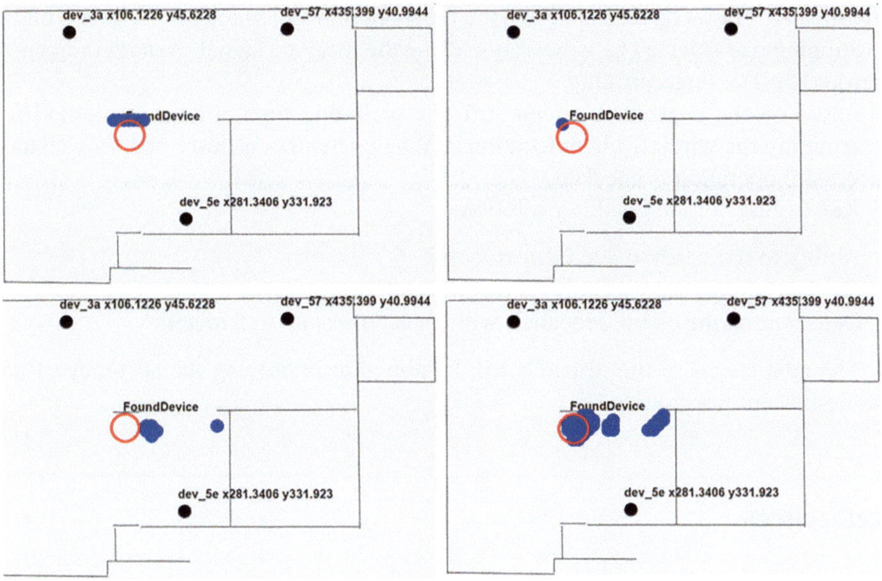

Fig. 11 Target device positioning results (actual device location circled)

5 Discussion

5.1 Operation Checking of the Wireless Client Location Method

To check the operation of the positioning method, the following experiment was carried out on the territory of the office premises. A device that supports IEEE 802.11 standards has been placed in this office. Sensors according to the map in Fig. 11 were set to the appropriate coordinates (points labeled as "dev_ ..."). At random times, the process of determining the placed device location was launched. In this way, several lists of coordinates were obtained.

Figure 11 shows the positioning results of this device. The actual location of this device is circled. It should be noted that the result of the algorithm is a list of the desired device locations, initially presented as a table. This table can be interpreted in a different graphical form that is understandable to the user, as in Fig. 11.

The location accuracy in this case was 1–5 m.

6 Conclusion

At the initial stage of the work, various positioning technologies in Wi-Fi networks, methods for their implementation, and their advantages and disadvantages were investigated. Based on data analysis, the most optimal technology for wireless client positioning was selected by trilateration using the received signal strength indication metric (RSSI) – fingerprinting.

Based on the analysis of theme articles, including foreign ones, a method for determining the wireless client location in Wi-Fi networks is described, as well as a successful operation is checked.

Key features of the resulting solution:

- Ability to use ready-made infrastructure.
- No dependency on a specific hardware vendor.
- Determining the client's location with an accuracy up to 5 meters.

The next stages of the research will be aimed at improving the accuracy of the wireless client positioning.

References

1. Drepa, V.E., Kistruga, A.Yu., Kovtsur, M.M., et al.: Investigation of fingerprinting method for IEEE 802.11 wireless client location. Notes of scientist, no. 3-2, pp. 137–141. EDN BKRUYE (2022)
2. Doynikova (Fedorchenko), E.V., Fedorchenko, A.V., Ushakov, I.A., et al.: Decision support for response to disturbances in control systems using graph models. In: International Scientific Conference on Control Problems in Technical Systems, vol. 1, pp. 278–282. EDN NUWIUQ (2021)
3. Bragin, D.S., Pospelova, I.V., Cherepanova, I.V., Serebryakova, V.N.: Radio-frequency technologies of local positioning in public health. Radioelectronics. **23**(3), 62–79 (2020). https://doi.org/10.32603/1993-8985-2020-23-3-62-79., EDN CPIBLZ
4. Rusanov, P.I., Yurochkin, A.G.: Analysis of basic positioning methods in Wi-fi networks. Bull. Voronezh Inst. High Technol. **1**(28), 100–102., EDN DXCJNB (2019)
5. Andreyev, R.A., Ostroumov, S.I., Fedorov, A.S.: Methods of positioning in Wi-Fi networks. Econ. Qual. Commun. Syst. **3**(21), 50–63. EDN QCWKAM (2021)
6. Nazarhudoev, M.D., Repinsky, V.N.: Methods of WI-FI network subscriber positioning and the possibility of improving their accuracy. REDS Telecommun. Devices Syst. **11**(4):32–41. EDN KYYMIB (2021)
7. Rea, M., Fakhreddine, A., Giustiniano, D., Lenders, V.: Filtering noisy 802.11 time-of-flight ranging measurements from commoditized WiFi radios. IEEE/ACM Trans. Networking. **25**(4), 2514–2527 (2017). https://doi.org/10.1109/TNET.2017.2700430
8. Kuchin, Y., Iksanov, S.S., Rozhdestvensky, S.K., Koryakov, A.N.: Development of a positioning and object control system using Wi-fi wireless technology. Sci. Bull. Novosibirsk State Techn. Univ. **3**(60), 130–146 (2015). https://doi.org/10.17212/1814-1196-2015-3-130-146., EDN UNEYOT
9. Bogurenko, P.A., Burlakov, M.E.: Review of methods of local positioning of objects in Wi-fi networks. Elect. Eng. Inf. Technol. Control Syst. **23**, 146–158., EDN ZJSOOJ (2017)
10. Hu, J.: Wireless industrial indoor localization and its application. UiT The Arctic University of Norway. Master thesis in Industrial Engineering (June 2017)
11. Haidarov, H., Kimizbaeva, A.E., Shukurova, A.A.: The system of local positioning of objects based on RTLS-technology. In: Actual Scientific Research in the Modern World, vol. 12-1(56), pp. 119–124. EDN WFWHBK (2019)

12. Reva, I.L., Bogdanov, A.A., Malakhova, E.A.: Application of Wi-fi access points for motion detection at an object. Sci. Bull. Novosibirsk State Tech. Univ. **3**(68), 104–125 (2017). https://doi.org/10.17212/1814-1196-2017-3-104-125., EDN XBUMNG
13. Swangmuang, N., Krishnamurthy, P.: An effective location fingerprint model for wireless indoor localization. Pervasive Mob. Comp. **4**(6), 836–850., ISSN 1574-1192 (2008)
14. Jacobsen, E.: Frequency Dependence in Free Space Propagation. DSPRelated. Accessed 5 Feb 2022. https://www.dsprelated.com/showarticle/62.php (2008)
15. «Robot Indoor Positioning and Navigation Based on Improved WiFi Location Fingerprint Positioning Algorithm». Hemin Ye and Jiansheng Peng. Accessed February 2022. https://www.hindawi.com/journals/wcmc/2022/8274455/
16. Kovtsur, M.M., Yurkin, D.V., Gerling, E.Yu., Gogol, R.A.A.: Security of wireless local area networks. Tutorial for term paper. The Bonch-Bruevich SPbSUT, Saint-Petersburg (2021). Accessed 1 Sept 2021
17. DuBois, P.: MySQL Cookbook: Solutions for Database Developers and Administrators, Third Edition. ISBN: 978-1-449-37402-0 (2014)
18. Targonskaya, A.I., Tsvetkov, A.Y.: Development of secure web-interface to control devices in network. In: Actual Problems of Infotelecommunications in Science and Education, vol. 1, pp. 734–739., EDN YYLVCT. The Bonch-Bruevich SPbSUT, Saint-Petersburg (2019)
19. Nikanorkina, O.V., Shibaikin, S.D.: Analysis of WI-FI signal watching. In: Proceedings of XXII scientific-practical conference of young scientists, graduate students and students, N.P. Ogarev National Research University of Mordovia: conference proceedings. In 3 parts, Saransk, September 25 – 01, 2018, pp. 154–157, EDN ZAZVEL (2019)
20. Wilson, R.: Propagation losses through common building materials 2.4 GHz vs 5 GHz, E10589, Magis Network (2002)
21. Friis, H.T.: A note on a simple transmission formula. Proc. IRE. **34**(5), 254–256 (1946). https://doi.org/10.1109/JRPROC.1946.234568
22. Li, G., Geng, E., Ye, Z., Xu, Y., Lin, J., Pang, Y.: Indoor positioning algorithm based on the improved RSSI distance model. Sensors. **18**(9), 2820 (2018). https://doi.org/10.3390/s18092820
23. Volkogonov, V.N., Kazantsev, A.A., Katasonov, A.I., Orlov, G.A.: Analysis of Wi-fi networks security. In: Actual Problems of Infotelecommunications in Science and Education, vol. 1, pp. 270–275., EDN JQALFF. The Bonch-Bruevich SPbSUT, Saint-Petersburg (2019)
24. Desnitskii, V.A., Sakharov, D.V., Chechulin, A.A., et al.: Information Protection in Data Centres, p. 92., EDN WLKSHP. The Bonch-Bruevich SPbSUT, Saint-Petersburg (2019)
25. Ushakov, I.A., Kotenko, I.V., Ovramenko, A.Y., et al.: Combined approach to detecting insiders in computer networks. Bulletin of St. Petersburg State University of Technology and Design. Series 1: Natural and Technical Sciences, vol. 4, pp. 66–71. https://doi.org/10.46418/2079-8199_2020_4_10. EDN HSQQIJ (2020)
26. Vasilishin, N.S., Ushakov, I.A., Kotenko, I.V.: Research of algorithms of network traffic analysis using big data technologies to detect computer attacks. In: Information Technology in Management: Proceedings of the 9th conference on management problems, Saint-Petersburg, pp. 670–675, EDN XFCHUR (2016)
27. Maxim, K., Anton, K., Anastasiya, M., Pavel, P., Vladimir, V.: Research of wireless network traffic analysis using big data processing technology. In: 2021 13th International Congress on Ultra Modern Telecommunications and Control Systems and Workshops (ICUMT), pp. 115–121. https://doi.org/10.1109/ICUMT54235.2021.9631682 (2021)
28. Khakimov, A., Muthanna, A., Kirichek, R., Koucheryavy, A., Muthanna, M.S.A.: Investigation of methods for remote control IoT-devices based on cloud platforms and different interaction protocols. In: 2017 IEEE Conference of Russian Young Researchers in Electrical and Electronic Engineering (EIConRus), pp. 160–163. https://doi.org/10.1109/EIConRus.2017.7910518 (2017)
29. Muthanna, A., Ateya, A., Khakimov, A., Gudkova, I., Abuarqoub, A., Samouylov, K., Koucheryavy, A.: Secure and reliable IoT networks using fog computing with software-defined networking and blockchain. J. Sens. Actuator Netw. **8**, 15 (2019)

30. Muthanna, A., Prokopiev, A., Paramonov, A., Koucheryavy, A.: Comparison of protocols for Ubiquitous wireless sensor network. In: 2014 6th International Congress on Ultra Modern Telecommunications and Control Systems and Workshops (ICUMT), pp. 334–337. https://doi.org/10.1109/ICUMT.2014.7002123 (2014)
31. Lovinger, N., et al.: Detection of wireless fake access points. In: 2020 12th International Congress on Ultra Modern Telecommunications and Control Systems and Workshops (ICUMT), 113–118. IEEE (2020)
32. Kovtsur, M., et al.: Investigation of attacks and methods of protection of wireless networks during authorization using the IEEE 802.1 x protocol. In: The 5th International Conference on Future Networks & Distributed Systems, 555–561 (2021)
33. Petrov, V.A., Kovtsur, M.M., Kistruga, A.Yu., Shterenberg, S.I.: Investigation of frame types used for positioning. In: Regional Informatics and Information Security: Proceedings of XII Saint-Petersburg Interregional Conference, Saint-Petersburg, pp. 62–67, EDN LNXTMY, 27–29 Nov 2021
34. Kovtsur, M.M., Peshkov, A.I., Mislivskiy, B.S., et al.: Software for the automatic acquisition of the IEEE 802.11 wireless networks traffic: no. 2021669199: application form. 26.11.2021 : publ. 07.12.2021. State Registration Certificate for Computer Software no. 2021680157 Russian Federation, Saint-Petersburg: The Bonch-Bruevich SPbSUT, EDN KLBBEG
35. Fedorova, A.E., Gerling, E.Yu., Ahrameeva, K.A., Andrianov, V.I.: Development of Web-interface for monitoring system of wireless networks of IEEE 802.11 family. In: Regional Informatics and Information Security: Proceedings of XII Saint-Petersburg Interregional Conference, Saint-Petersburg, pp. 381–385, EDN DXLBVX, 27–29 Nov 2021
36. Huang, Y., Zheng, J., Xiao, Y., Peng, M.: Robust localization algorithm based on the RSSI ranging scope. Int. J. Distrib. Sensor Netw. **11**(2) (2015). https://doi.org/10.1155/2015/587318

Index

A. A. Abd El-Latif et al. (eds.), *Secure Edge and Fog Computing Enabled AI for IoT and Smart Cities*, EAI/Springer Innovations in Communication and Computing, https://doi.org/10.1007/978-3-031-51097-7

MIX
Papier aus verantwortungsvollen Quellen
Paper from responsible sources
FSC® C105338

If you have any concerns about our products,
you can contact us on
ProductSafety@springernature.com

In case Publisher is established outside the EU,
the EU authorized representative is:
Springer Nature Customer Service Center GmbH
Europaplatz 3, 69115 Heidelberg, Germany

Printed by Libri Plureos GmbH
in Hamburg, Germany